Functional Dimensions of Ape-Human Discourse

Functional Dimensions of Ape–Human Discourse

Edited by
James D. Benson and
William S. Greaves

LONDON   OAKVILLE

Published by

Equinox Publishing Ltd.
UK: Unit 6, The Village, 101 Amies Street, London, SW11 2JW
USA: DBBC, 28 Main Street, Oakville, CT 06779
www.equinoxpub.com

First published 2005

British Library Cataloguing-in-Publication Data
A catalogue record for this book is available from the British Library.

ISBN 1-904768-05-9 (Hardback)

Library of Congress Cataloging-in-Publication Data

Functional dimensions of ape-human discourse / edited by James D. Benson
and William S. Greaves.
     p. cm. -- (Functional linguistics)
 Includes bibliographical references and index.
 ISBN 1-904768-05-9 (hb)
 1. Bonobo--Psychology. 2. Human-animal communication. I. Benson,
James D. II. Greaves, William S. III. Series.
  QL737.P96F86 2004
  156'.36--dc22

                                        2004008309

Typeset by Catchline, Milton Keynes (www.catchline.com)
Printed and bound in Great Britain by Antony Rowe Ltd.,
Chippenham, Wiltshire

Contents

List of contributors

James D. Benson is from Glendon College, York University.
Meena Debashish is an independent scholar.
Peter H. Fries is from Central Michigan University.
William S. Greaves is from Glendon College, York University.
Kazuyoshi Iwamoto is from York University.
Jennifer Lukas is from York University.
Michael O'Donnell is from WagSoft Systems.
Sue Savage-Rumbaugh is from the Great Ape Trust of Iowa.
Jared P. Taglialatela is from the Yerkes National Primate
    Research Center, Emory University.
Paul J. Thibault is from Agder University College.

Preface

This volume brings together for the first time two strands of research: primatology and functional linguistics, both of which have a long history. Systemic Functional Linguistics brings to ape-language research a model of language-in-context along two axes: metafunction, with three equally important modes of meaning (ideational, interpersonal, and textual); and stratification, with five correlated strata (context, semantics, lexicogrammar, phonology, and phonetics) (Firth, 1957; Halliday, 1995; Halliday, 1994; Halliday, 1978; Halliday, 1975; Halliday, 1967; Halliday & Hasan, 1985; Halliday & Matthiessen, 1999; Matthiessen, 1995; Hasan, 1985). Sue Savage-Rumbaugh brings to Systemic Functional Linguistics a vast amount of rich language-in-context data (Rumbaugh & Washburn, 2003; Savage-Rumbaugh et al., 1993; Savage-Rumbaugh & Lewin, 1994; Savage-Rumbaugh, Shanker & Taylor, 1998; Savage-Rumbaugh et al., 2001; Taglialatela et al., 2004).

The common thread of all four chapters is ape-human discourse interaction. Chapter 1 provides evidence for the bonobo Kanzi's control of interpersonal discourse semantics in a complex negotiation with Sue Savage-Rumbaugh. By making his initial command ('GO OPEN GROUPROOM'), Kanzi assigned the complementary discourse role of compliance to Sue. In this particular negotiation, Kanzi used graphic symbols called lexigrams, rather than vocalizations, as his primary medium of communication. Chapter 2 reinterprets the experimentation documented in Savage-Rumbaugh et al. (1993) in terms of ideational lexicogrammar, and provides corroborative evidence for Kanzi having some degree of higher-order consciousness. Although discourse is not the focus of the discussion, the 660 novel requests that constituted the experiment nevertheless took place in a discourse context: the experimenter made a series of commands, and assigned Kanzi the complementary discourse role of compliance. Since compliance required that Kanzi perform actions, neither lexigrams nor vocalizations were at issue.

Chapter 3, with a focus on the evolution of language, provides evidence for the bonobo Panbanisha's control of interpersonal discourse semantics in another complex negotiation with Sue Savage-Rumbaugh (based on Panbanisha's understanding of English lexicogrammar) where the difference between talking *about* Kanzi and talking *to* Kanzi is expressed by Panbanisha's timing and choice of vocalizations. In this case, Sue is asking questions, which assign Panbanisha the complementary role of answer, and Panbanisha's primary medium of expression is vocalization, with proximal vocalizations addressed to Sue, and distal vocalizations addressed to Kanzi in an adjoining location.

Chapter 4 addresses Kanzi's proximal vocalizations in greater detail, by considering whether or not a hearer can reasonably interpret Kanzi's vocalizations as a dialect of spoken English. We develop a methodology for identifying Kanzi's vocalizations as allophones of English phonemes by making explicit the differing contributions to meaning-making made at all strata of language-in-context.

Acknowledgements

The material in Chapter 1 was originally presented at the First International Workshop of the Systemic Functional Research Community on Interpersonal and Ideational Grammar at the University of Leuven. The material in Chapter 2 was originally presented at the Second International Workshop of the Systemic Functional Research Community on Interpersonal and Ideational Grammar at the University of Leuven. The material in Chapter 3 was originally presented at the Harvard Evolution of Language Conference. The material in Chapter 4 was originally presented at the 'mind AND world' conference 'Integrational Linguistics and Distributed Cognition' at the University of Natal. We wish to acknowledge Glendon College, York University, and the Fund for Scientific Research – Flanders (Belgium) for travel support.

Chapter 1 was previously published as 'Confrontation and support in bonobo-human discourse' in *Functions of Language*, 9, 1, 2002, pp. 1–38. We are grateful to John Benjamins Publishing Company, Amsterdam/Philadelphia (www.benjamins.com) for permission to republish it here.

Chapter 2 was previously published as 'Evidence for symbolic language processing in a bonobo (Pan paniscus)' in *Journal of Consciousness Studies*, 9,12, 2002, pp. 33–56. We are grateful to Imprint Academic (www.imprint.co.uk/jcs.html) for permission to republish it here.

Chapter 4 was previously published as 'Mind and brain in apes: a methodology for phonemic analysis of vocalizations of language competent bonobos' in *Language Sciences*, 26, 2004, pp. 643–60. We are grateful to Elsevier Science (www.elsevier.com) for permission to republish it here.

We are very much indebted to our co-authors, particularly for their patience. We are particularly grateful to Jennifer Lukas for editorial assistance. In addition, we want also to acknowledge the contributions of students and colleagues: Chris Cleirigh, Terry Deacon, Bill Downes, Rhondda Fahey, Nan Fries, Michael Halliday, Ruqaiya Hasan, Grace Hunt, Naomi Knight, Richard Lewontin, Mary Minehan, Ann-Louise Macintyre, Erin O'Hara, Shane Lukas, Severine Renard, Dan Rice, Duane Rumbaugh, Vicky Saucier, Lieven Vandelanotte and Geoff Williams.

References

Firth, J. R. (1957) *Papers in Linguistics 1934–1951*. London: Oxford University Press.

Halliday, M. A. K. (1995) On language in relation to the evolution of human consciousness. In S. Allén (ed.) *Of Thoughts and Words*: *proceedings of the Nobel symposium 92*: *the relation between language and mind* 45–84. Singapore: Imperial College Press.

Halliday, M. A. K. (1994) *An Introduction to Functional Grammar*. London: Edward Arnold.

Halliday, M. A. K. (1978) *Language as Social Semiotic*: *the social interpretation of language and meaning*. London: Edward Arnold.

Halliday, M. A. K. (1975) *Learning How to Mean*: *explorations in the development of language*. London: Edward Arnold.

Halliday, M. A. K. (1967) *Intonation and Grammar in British English*. The Hague: Mouton.

Halliday, M. A. K. and Hasan, R. (1985) *Language, Context and Text*: *a social semiotic perspective*. Language and Learning Series. Geelong, Vic: Deakin University Press.

Halliday, M. A. K. and Matthiessen, C. (1999) *Construing Experience through Meaning*: *a language-based approach to cognition*. London and New York: Cassell.

Hasan, R. (1985) Meaning, context and text: fifty years after Malinowski. In J. Benson and W. Greaves (eds) *Systemic Perspectives on Discourse*: *selected applied papers from the ninth international systemic workshop vol. 2*. Norwood, N.J.: Ablex Publishing Corporation.

Matthiessen, C. (1995) *Lexicogrammatical Cartography*: *English systems*. Tokyo: International Language Sciences Publishers.

Rumbaugh, D. M. and Washburn, D. A. (2003) *Intelligence of Apes and Other Rational Beings*. New Haven and London: Yale University Press.

Savage-Rumbaugh, S., Murphy, J., Sevcik, R., Brakke, K., Williams, S. and Rumbaugh, D. (1993) *Language Comprehension in Ape and Child*. Monographs of the Society for Research in Child Development. Chicago: University of Chicago Press.

Savage-Rumbaugh, S. and Lewin, R. (1994) *Kanzi*: *an ape at the brink of the human mind*. New York: Wiley.

Savage-Rumbaugh, S., Shanker, S. and Taylor, T. (1998) *Apes, Language and the Human Mind*. New York: Oxford University Press.

Savage-Rumbaugh, S., Fields, W. M. and Taglialatela, J. P. (2001) Language, speech, tools and writing: a cultural imperative. *Journal of Consciousness Studies* 8(5–7): 273–92.

Taglialatela, J., Rumbaugh, D., Savage-Rumbaugh, S., Benson, J. and Greaves, W. (2004) Language, apes and meaning-making. In G. Williams and A. Lukin (eds) *The Development of Language*: *functional perspectives on species and individuals*. London: Continuum.

1 The interpersonal dimension: confrontation and support in bonobo-human discourse

James D. Benson, Peter Fries, William S. Greaves, Kazuyoshi Iwamoto, Sue Savage-Rumbaugh and Jared P. Taglialatela

Abstract

As part of a program to explore the communicative abilities of bonobo apes within the human-ape culture at the Language Research Center at Georgia State University, we made two complementary analyses of a conversation between Sue Savage-Rumbaugh and Kanzi. We made both a conversation analysis and a lexicogrammatical analysis of their interaction. The conversation analysis revealed the participants negotiating the interpersonal meanings of confrontation and support, while the lexicogrammatical analysis revealed the ideational domain of the confrontation and support. Although many of the contributions of both participants did not fully express all the relevant meanings, both participants interpreted each other's contributions in terms of their relevance to the patterns of interpersonal and ideational meanings being expressed in the conversation. We conclude that Kanzi's considerable language abilities have been underestimated. First, Kanzi (despite his limited syntax) and Sue jointly construe a recognizable social world through discourse. Second, in exchanging discourse roles with Sue, Kanzi negotiates the asymmetrical power relation between himself and Sue. Finally, Kanzi's accomplishment suggests that discourse semantics is a powerful motivation for the evolution of both interpersonal and ideational grammar.

1 Introduction

Ape language research has a history of controversy (Savage-Rumbaugh & Lewin, 1994; Pinker, 1994; Deacon, 1997). In this paper, we propose a new approach to Kanzi's communicative abilities. We contextualize this approach with a brief look at one of the more controversial issues.

1.1 Conflicting criteria for language: syntax vs. semiosis

At the present time there are conflicting criteria for determining whether apes use language. Is the defining criterion for language the presence of syntax or the presence of semiosis?

Calvin and Bickerton (2000), for example, claim that without syntax there can be only structureless protolanguage. Deacon (1997), on the other hand, claims that the key to language is the ability to symbolize. These two conflicting criteria for language arrive at dramatically different results when judging Kanzi's behaviour.

Bickerton (Calvin & Bickerton, 2000: 38–40) denies Kanzi the ability to understand a novel sentence like 'Kanzi, go to the office and bring back the red ball' as human language, even if he goes to the office and brings back the red ball, because Kanzi has only a protolanguage without syntax:

> In fact, in understanding what something means, we have all sorts of clues from semantics, pragmatics, and situational context that are quite useless when it comes to production. I don't think for a moment that Kanzi knew the grammatical structure of 'Go to the office and bring back the red ball' – if he knew what 'go,' 'office,' 'bring,' and 'red ball' meant, he wouldn't have to be a rocket scientist to figure out what he had to do.

At the same time, Bickerton (Calvin & Bickerton, 2000: 23–4), referring to Deacon (1997), rejects a semiotic account of Kanzi's language abilities:

> I'd like to comment on a recent suggestion that the rubicon between our species and others falls at the symbolic rather than syntactic level. In other words, it's words, not sentences, that dramatically distinguish our species from others.... In fact, as was apparent nearly two decades ago, the real rubicon, unpalatable though this may be to the philosophically minded, is syntax, not symbols.

Deacon (1997: 124–6) indeed credits Kanzi with symbolic language, but does not, as Bickerton would suggest, limit symbolic value to words. Nor does he privilege syntax. What makes meaning symbolic is the possibility of systematic choices. Although sentence internal grammar is not our focus, we see Bickerton's 'syntax' as one of the ways in which systematic choice manifests itself in recognizable structures. But 'syntax' is only one of the structural resources available. There are also structures of intonation and structural configurations in semantics. And discourse itself is structured. All of these structures do the work of making and expressing symbolic meaning. As Deacon (1997: 312) points out, the right hemisphere of the brain is heavily involved in such processing:

> The right hemisphere is not the non-language hemisphere. It is critical
> for the large-scale, semantic processing of language, not word meaning
> so much as the larger scale symbolic constructions that words and
> sentences contribute to: complex ideas, description, narratives, and
> arguments. Symbol construction and analysis do not end with the
> end of a sentence, but in many regards begin there. The real power of
> symbolic communication lies in its creative and constructive power.
> Since symbolic representation is intrinsically compositional, there
> is no upper bound to the compositional complexity of a symbolic
> representation.

In our study we are particularly concerned with one of these 'larger scale symbolic constructions that words and sentences contribute to', i.e. casual conversation. Indeed, our study, by examining Kanzi's discourse, indicates that Bickerton's assumption (Calvin & Bickerton, 2000: 137) 'that language began in the form of a structureless protolanguage, something like an early stage pidgin, without any formal structure – just handfuls of words strung together' fails to account for the interactional dimension of Kanzi's communicative abilities. In our view, there is a larger structure – interactional structure – which Kanzi controls even while his control of grammatical structure may be debatable.

1.2 How Kanzi learned

The most obvious obstacle to overcome when studying the actual or potential language abilities of apes is that their vocal tract is such that they cannot produce the speech sounds that humans do. The solution to this problem by ape language researchers at the Language Research Center at Georgia State University is a board of abstract symbols, or lexigrams, which apes use as words to speak with. In the early 1980s, as part of an ongoing research program aimed at uncovering the ape's potential for language acquisition, a female bonobo, Matata, was being trained, with limited success, to recognize the meanings of an inventory of lexigrams. Matata's adopted son, Kanzi, was

> present during Matata's training, but was thought to be too young to learn, and so was virtually ignored during these sessions. However, to everyone's astonishment, Kanzi began using the lexigram board himself, demonstrating that he had learned the meaning of a number of lexigrams without explicit teaching. As a result, Kanzi has captured the attention of scientists outside the ape language research community (Savage-Rumbaugh & Lewin, 1994; Savage-Rumbaugh et al., 1998).

1.3 Casual conversation: fundamental principles

Although Kanzi's language abilities have been extensively studied in experimental settings, much of his learning to interact with humans takes place through the mode of natural conversation. The discourse that we focus on in this paper is a typical interaction between Sue and Kanzi, which allows us to examine the world that is constructed in this conversation.

In order to examine this jointly constructed world, we have made extensive use of the analytical techniques of Eggins and Slade (1997), whose research aim was 'to explore what it means to claim that casual conversation is critical to the social construction of reality'. Their goal was to uncover the general principles underlying implicitness of casual conversation:

> to show how social structures are negotiated, how attitudes and values shaped by differences of concern to the institutionalized social context are reflected in and modified by casual talk (1997: 316).

They found two principles at work. The first principle is that casual conversation is only apparently casual. On the one hand, 'the social functions of casual talk remain largely invisible to its participants' (Eggins

& Slade, 1997: 316), but on the other hand, casual talk maintains and constructs reality. This is the 'paradox' of casual conversation (Eggins & Slade, 1997: 17):

> we experience casual conversation as probably the only context in which we are talking in a relaxed, spontaneous and unselfconscious way. We feel it is the only place where we are really free to be ourselves and yet, at the same time, we are hardly free at all. We are in fact very busy reflecting and constituting our social world.

In other words, this paradox is the result of:

> a tension between, on the one hand, establishing solidarity through the confirmation of similarities, and, on the other, asserting autonomy through the exploration of differences. (Eggins & Slade, 1997: 22)

The second principle is that the exchange of interpersonal meanings is the engine that drives discourse, first because 'the primary task of casual conversation is the negotiation of social identity and social relations', and second because 'the open-ended, turn-taking organization of conversation differentiates it from other linguistic activities' (Eggins & Slade, 1997: 49–50).

As a result, discourse foregrounds interpersonal, rather than ideational or textual meanings. For example, 'the observation that anything can be the topic in casual talk in casual conversation… suggests that the important work of casual conversation is not in the exploration of ideational meanings'. Rather, ideational meanings play a facilitating role: 'any ideational domain (or Field) serves as the environment for the exploration of social similarities and differences' (Eggins & Slade, 1997: 50).

2   Questions to be asked

An examination of Kanzi's discourse offers a new perspective on Kanzi's linguistic abilities. The embedding of Kanzi's utterances in the flow of discourse reveals that his communicative abilities may have been underestimated. Unlike the study of Kanzi's utterances in terms of decontextualized syntax, consideration of the interactive dimension of discourse allows us to ask three inter-related questions, which have both ontogenetic and phylogenetic implications.

1) What kind of social world are Kanzi and Sue jointly construing?

As we shall see, this world is not random and anti-social, but highly organized and social. Moreover, the turn-taking principle on which it is based is the environment *par excellence* for the development of 'role-reversal imitation', which is crucial in the ontogeny of human cognition (Tomasello, 1999: 103–7).

2) To what extent does Kanzi participate in the human, continuous negotiation of differential power relations?

As we shall also see, Kanzi fully engages Sue in an activity fundamental to the development of human cognition, described by Tomasello (1999: 170):

> sometimes the semantic content of the discourse, what is being talked about over multiple discourse turns, expresses differing and sometimes conflicting construals of things.

The point is that Kanzi and Sue are dealing with conflict by jointly construing discourse, rather than by hurling sticks and stones at each other.

3) To what extent might discourse be the motivational environment for the development of both interpersonal grammar (mood) and ideational grammar (processes, participants, and circumstances)?

The fact that Kanzi's moves in the first exchange, a negotiation of his demand for goods and services, are more richly patterned than his grammatical structures, is suggestive. Halliday's (1994: 70) perspective on human ontogeny is this:

> in the life history of an individual child, the exchange of goods-&-services, with language as the means, comes much earlier than the exchange of information: infants typically begin to use linguistic symbols to make commands and offers at the age of about nine months, whereas it may be as much as nine months to a year after that before they really learn to make statements and questions…. It is quite likely that the same sequence of developments took place in the early evolution of language in the human race, although that is something we can never know for certain.

When Kanzi demands goods and services in a human language-rich environment, it does not mean that his behaviour represents a normal stage of evolutionary development. It *does* say, however, that this behaviour (as opposed to the social evolution of grammar) is not unique to humans. Ontogenetically it is interesting that Kanzi's exchange pattern deals predominantly with goods and services: a pattern of communicative behaviour familiar to us because we know what human children do (Halliday, 1975; Painter, 1984: 252; 1999: 181–2). It is also interesting to speculate as to whether or not communication to achieve the exchange of goods and services formed an early basis for further differentiation (Halliday, 1994: 70) in the course of the evolution of human language (cf. Matthiessen, 2004).

3   Data overview

3.1 Context of culture

For the past two decades, Kanzi has been, and continues to be, the subject of a study on non-human language acquisition. It has been demonstrated that Kanzi possesses a sophisticated comprehension of spoken English, as well as the ability to produce novel utterances via a lexigram keyboard (Savage-Rumbaugh et al., 1998; Savage-Rumbaugh & Lewin, 1994). While these studies aimed at evaluating Kanzi's language abilities were conducted within the confines of the laboratory in highly controlled situations, Kanzi's acquisition and application of language takes place in a very different context.

As an adult, Kanzi's linguistic competencies reflect the social environment that he inhabits. That is, his daily interactions with researchers and caregivers at the Language Research Center are mediated by spoken English, gestures, and vocalizations, as well as lexigram keyboard use. Humans make requests of Kanzi, and he in turn is able to respond with utterances of his own. In addition, Kanzi utilizes lexigrams to alter his environment (make food requests, go outdoors, watch television, etc.).

Kanzi is confined to indoor and outdoor enclosures, and therefore his linguistic utterances are often directed towards researchers that control the food and housing areas he has access to. As one can imagine, not all of Kanzi's requests can be met. His diet is limited so as to provide him with balanced nutrition. Kanzi must move from one cage to another for testing, cage-cleaning or repairs, and often must remain in certain areas while research is conducted with other apes at the lab.

Just as researchers and caregivers must deny some of Kanzi's requests, Kanzi also, on occasion, refuses to do what he is asked. Given his size and strength, it is nearly impossible to physically force Kanzi to do something. Therefore, a degree of negotiation underlies many of the linguistic interactions between Kanzi and his human counterparts.

It is this cultural context that provides the framework for Kanzi's linguistic interactions with humans. Requests made of, or by, Kanzi are often met with alternative suggestions, conditional compliance, or counter-requests. The interspecies dialogue, therefore, is marked by confrontation and support, as both Kanzi and the researchers that work with him negotiate their common world through language.

3.2 Data

The data for analysis consists of an interaction between Kanzi and Sue Savage-Rumbaugh, recorded on videotape. Prior to the interaction, Sue has left the Middle Test Room with Nyota (a baby bonobo), to get a ball that Kanzi has requested. Sue and Nyota return after a few minutes. During the next minute of silence, Sue sets up the lexigram board for Kanzi so that it is usable, and visible to the camera, and gives the ball to Nyota, who holds it briefly, and then drops on the floor, making a loud noise. Kanzi then turns to the lexigram board and points to the symbols GO OPEN GROUPROOM. A transcript of this discourse is displayed below.

Figure 1: 'go open grouproom' transcript

Key: K = Kanzi
S = Sue
UPPER CASE = the lexigrams that Kanzi points to
Initial Capitals = When Sue is pointing to lexigrams in addition to saying them
lower case = When Sue is speaking without pointing to lexigrams
Arabic numeral followed by / = Exchanges
Arabic numeral followed by / = Turns
Lower case letters = Moves

Caps: arabic/arabic/lower case = exchange/turn/move

Note: Exchange, turn, and move are the constituent units of discourse for Eggins and Slade. A turn
consists of one or more moves, and is bounded by change of speaker.
An exchange begins with an opening move, proceeds through a variable number of turns,
and is terminated by a new opening move (whether or not by the same speaker).

(K1/1)   GO OPEN GROUPROOM

(S1/1)   go open grouproom

(K1/2)   [gestures toward the grouproom]

(S1/3/a)   oh over there

(S1/3/b)   that's that would be a fun thing to do

(S1/3/c)   Yes Sue Wants To Open Grouproom

(S1/3/d)   but Grouproom Is Broken

(K1/4)   BROKEN

(S1/5/a)   yeah it's broken it's broken

(S1/5/b)   it's broken in there because mike is working on something

(K1/6)   PLAYYARD

(S1/7)   playyard

(K1/8)   [vocalizing]

(S1/9/a)   oh

(S1/9/b)   well Yes Playyard… Tomorrow

(S1/9/c)   we can't go to the playyard today because we have to go through
the grouproom

(S1/9/d)   and the grouproom is broken

(S1/9/e)   Today We Will Play Here

(S1/9/f)   In Middletestroom

(K1/10)   [points toward MIDDLETESTROOM]

(S1/11/a) yeah yes yes middletestroom mm hmmm

(S1/11/b) mm hmm that's what we're going to do

(K1/12)   BEDROOM

(S1/13)   and you're pointing are you pointing to the Bedroom

(K1/14)   vocalization/BEDROOM

(S1/15)   the bedroom

(K1/16)   [vocalization – nods head up and down]

(S1/17)   you want to Open Bedroom

(K1/18)   [BEDROOM – K uses lexical info to provide yes/no info]

(S1/19/a) well kanzi There Are Toys In The Bedroom For You

(S1/19/b) We Can Play with those Toys in the Middletestroom

(S1/19/c) Kanzi Is Staying In Middletestroom Today

(S1/19/d) because mr. ida is Grabbing your Picture with his Camera

(S1/19/e) uh huh that's why we're staying in here

(S1/19/f) mm hmm we can do things in here

(K1/20)   [Kanzi gestures toward the bedroom with his left hand]

(S2/1)   What Do You Want Me To Get

(K2/2)   SURPRISE

3.2.1 *Kanzi's gestures and vocalizations*

Kanzi often moved or vocalized throughout the conversation. We have coded Kanzi's gestures and vocalizations only when they were clearly relevant to the negotiation. This occurred when Sue clearly interpreted them and reacted to them as part of the conversation.

4 Analysis

We have used two complementary types of analysis. The first examines the discourse moves in the interaction to see how the participants interacted; the second examines the lexicogrammatical patterns in the text to discover the point of the dispute.

4.1 The interaction

4.1.1 *Theory: speech functions*

The technique of analysis is the Eggins and Slade network of speech functions displayed in Figure 2.

The network allows us to track interactive discourse in terms of *exchanges*, *turns*, and *moves*. The network is a complex of symbolic systems. A system consists of a set of terms or features which share an entry condition. (In this section, the terms and entry conditions of systems are printed in italics.) For example, the terms *open* and *sustain* share the entry condition *move*. Further differentiation of speech functions is achieved by making the terms of a system the entry condition of a new system. For example, the term *open* in the *move* system becomes the entry condition for a new system whose terms are *attend* and *initiate*. The network of systems is based on the concept that meaning is the product of choice among alternatives. Unless such choices are made systematically, there can be no meaning.

The first choice for the speaker in the network is between an *opening move* and a *sustaining move*. This captures the fundamentals of discourse: the need to start it going in the first place, and the need to keep it going once it is started.

There are two ways to *open* a discourse. The choice *attend* means to make a bid for the focus of an interactant's attention. The choice *initiate* means to set up a proposition or proposal for negotiation. At this point the network incorporates the Hallidayan system of basic speech functions (Halliday, 1994). An *offer* is a proposal giving goods and services, and a *command* is a proposal demanding goods and services. Propositions deal with information. A *statement* is a proposition giving information, and a *question* is a proposition demanding information.

These four basic speech functions are classified as *opening moves* by Eggins and Slade. The rest of the network develops systems of speech function arising from the initial choice of the *sustain* option rather than the *open* option. Since there are many ways to keep discourse going once it has begun (as well as many ways of bringing it to completion), it is not surprising that the choices flowing from the *sustain* option are more complex than those flowing from the *open* option.

There are two ways to *sustain* the proposition or proposal under negotiation; it can be *continued* or *reacted to*. *Continue* is the option for the same speaker, and *react* is the option for a different speaker.

If a speaker is *reacting*, two choices open up: *respond* or *rejoinder*. The choice of *respond* is a signal that the *exchange* is moving to completion, whether through *support* or *confrontation*: e.g. a *question* will be *answered* (or not), an *offer* will be *accepted* (or not), a *statement* will be *acknowledged* (or not), a *command* will be *complied with* (or not).

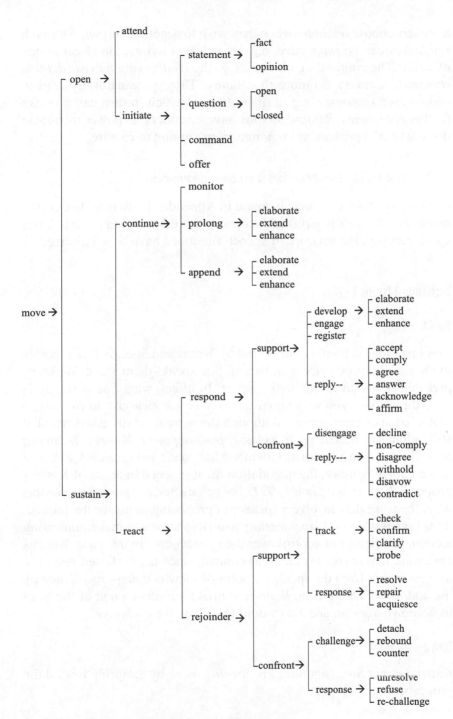

Figure 2: Speech function networks (Eggins & Slade, 1997)

Speakers choose *rejoinder* when they wish to negotiate further. Although *rejoinders* can be *supportive*, e.g. by *tracking moves*, the effect is one of delay. The *confronting* choices, e.g. the *challenging moves* (*detach, rebound, counter*), do more than delay. They quite naturally lead to *confronting responses*, e.g. a *re-challenge*, which in turn can provoke further *challenges*. *Rejoinders* thus have considerable power to *sustain* discourse, and prevent an *exchange* from coming to closure.

4.1.2   First analysis: speech functions in GO OPEN GROUPROOM

Our coding of the discourse is found in Appendix 1. We now discuss the results *turn* by *turn*, in terms of the meaning of the *moves* in context. Once again, terms in the network of speech functions have been italicized.

Exchange 1 (turns 1–20)

Turn 1

The first *move* is jointly constructed by Kanzi and Sue. As Kanzi points to the lexigrams GO OPEN GROUPROOM, Sue speaks them aloud. We interpret this as a kind of minimal support. In effect, what Sue is saying is something like: 'you've pointed to GO OPEN GROUPROOM; in my words that's 'go open grouproom''. Although the *move* is jointly constructed, it still belongs to Kanzi, and is a classic *opening move*. Kanzi is *initiating* an *exchange* with a *command* (demand for goods and services). Because it is an *opening move*, the negotiation must proceed in terms of Kanzi's proposal. Eggins and Slade (1997: 194) characterize *opening moves* this way: 'because they involve a speaker in proposing terms for the interaction, [*opening moves*] are generally assertive moves to make, indicating a claim to a degree of control over the interaction'. In this case, Kanzi's *command* maximizes his claim to control, since the preferred response to a *command* from the speaker's point of view is *doing-compliance* by the addressee. *Doing-compliance* requires more than a nod of the head in *acknowledgment*, and also would terminate the *exchange*.

Turn 2

Kanzi, but not Sue, *continues* his *opening move* by gesturing toward the grouproom.

Turn 3

Although Sue's first independent *move* is a *supportive* one, she is not yet in *compliance*. When she says 'oh over there' (3/a), she is *registering* what Kanzi has just said. Eggins and Slade (1997: 204) say that *registering moves*

> are reactions which provide supportive encouragement for the other speaker to take another turn. They do not introduce any new material for negotiation, and they carry the strong expectation that the immediately prior speaker will be the next speaker.

Sue's next three *moves* (3/b–d) represent a highly complex coding of ambivalence culminating in a *challenge* to Kanzi's *opening* proposal. We can tease this complexity apart by looking at the *moves* in sequence. On the face of it, Sue's 'that would be a fun thing to do' (3/b) is *supportive*, for three reasons. First, there is the positive note struck by the appraisal item 'fun', in 'a fun thing to do'. Second, it is a *responding reaction*, the function of which is to 'move the exchange toward completion' (Eggins & Slade, 1997: 200), and in this case, completion would be to *comply* (or not) with the *command*. Third, as a *developing move*, it indicates 'a very high level of acceptance of the previous speaker's proposition' (Eggins & Slade, 1997: 202). At the same time, however, there is a hint of trouble for Kanzi. The 'be' of 'that would be a fun thing to do' has been modalized by 'would', which places the doing in a probabilistic world, which lies between a definite yes (*compliance*) or a definite no (*non-compliance*) (Halliday, 1994: 356–7).

Sue's next *move* 'Yes Sue Wants To Open Grouproom' (3/c) elaborates the *supportive developing move* (3/b), but also separates 'wanting' from 'doing', i.e. wanting does not imply actually doing (cf. Halliday, 1994: 284–5). And because Sue speaks the 'Yes' as a separate tone group with falling-rising intonation, which indicates reservation, the stage is set for a reversal of some kind.

The reversal comes in the final *move* in this *turn* 'but Grouproom Is Broken' (3/d). Although (3/d) returns to the doing-domain of Kanzi's initial *command*, the reality presented amounts to *non-compliance*. But (3/d) is not the choice *confront: reply: non-comply* from the *respond* system. It clearly realizes a *confrontational rejoinder*. Since Sue doesn't offer any alternative, the next step is up to Kanzi. This is the essence of a *rebound*, which sends 'the interaction back to the first speaker, by questioning the relevance, legitimacy or veracity of another speaker's move' (Eggins & Slade, 1997: 212).

Turn 4

Kanzi's BROKEN (4) is a *register*, similar to Sue's (3/a). It is not a *tracking move*, such as a request for *confirmation*, since Kanzi looks away at this point, rather than looking questioningly at Sue.

Turn 5

Sue, however, treats Kanzi's BROKEN as a request for *confirmation*, and *resolves* the issue in (5/a) 'yeah it's broken it's broken'. Sue's *enhancement* of the *confirmation* of (5/a) with a 'reason' for the grouproom being broken, 'it's broken in there because mike is working on something' (5/b), serves only to emphasize her *rebounding move* (3/d). In other words, now that the clarification which Kanzi didn't request has been emphatically cleared up, the ball remains where it was, so to speak – in Kanzi's court.

Turn 6

At this point, Kanzi, unlike Sue, proposes an alternative to the grouproom scenario: PLAYYARD (6). This is the essence of another type of *challenging move*, the *counter*: 'these moves express confrontation by offering an alternative, counter-position or counter-interpretation of a situation raised by a previous speaker' (Eggins & Slade, 1997: 212). The nub of the proposal GO OPEN is presupposed, and very much still under negotiation, as far as Kanzi is concerned.

Turn 7

Sue's 'playyard' (7) is clearly a request for *confirmation*, since it is spoken with rising intonation, unlike the jointly constructed *opening move* (1), which was spoken with falling intonation.

Turn 8

The most we can say here is that Kanzi's vocalizing (8) is a *sustaining move*.

Turn 9

Sue's 'oh' (9/a) is a *registering move*, since Sue seems to be treating Kanzi's vocalization (8) as supplying positive polar information in *response* to her request for *confirmation* (7). This is Sue's second extended *turn*, but unlike *turn* 3, Sue's *challenging move* comes near the beginning, 'well yes play yard tomorrow' (9/b), which is then prolonged in *moves* (9/c–f). Because she is proposing a temporal alternative to Kanzi's *counter* in (5), Sue's *move* (9/b) is a *counter* rather than a *rebound*, as in (3/d). Sue *continues* her *counter* for four moves, by alternating reasons in support of her position with restatements of it. The *enhancing move* 'we can't go to the playyard today because we have to go through the grouproom' (9/c) is followed by the *extending move* 'and the grouproom is broken' (9/d); and the *enhancing move* 'Today We Will Play Here' (9/e) is followed by the *elaborating move* 'In Middletestroom' (9/f).

Turn 10

Kanzi points toward MIDDLETESTROOM, which is a *registering move*.

Turn 11

Sue's 'yeah yes yes middletestroom mm hmm' (11/a) treats Kanzi's (10) as a *move* needing an *affirmative reply*, and in 'mm hmm that's what we're going to do' (11/b), she *elaborates* the *counter* with which she began her previous *turn* once again.

Turn 12

Kanzi has become caught up in the discourse logic of *confrontation* (Eggins & Slade, 1997: 212):

> challenging moves more or less obligate the prior speaker to respond. However, very often the responses are themselves confronting: a query cannot be resolved, a counter is refuted, or a re-challenge launched. Reactions to rejoinder moves have themselves a certain rejoinder quality about them, and often lead to further challenging or tracking. It is not uncommon in casual talk to find lengthy sequences of such talk.

In Exchange 1 so far, Sue *reacted* to Kanzi's *opening* proposal, GO OPEN GROUPROOM, with a *challenging rejoinder*, 'but Grouproom Is Broken'

(3/d); Kanzi *reacted* to (3/d) with a *challenging rejoinder* of his own, PLAYYARD (6); and Sue *reacted* to (6) with another *challenging rejoinder*, 'well Yes Playyard Tomorrow' (9/b). At this point, Kanzi is still committed to the nub of his original proposal, the presupposed GO OPEN, and offers a second alternative location in yet another *challenging rejoinder* BEDROOM (12).

Turn 13

Sue *reacts* with a *tracking move* requesting *confirmation* 'and you're pointing are you pointing to the Bedroom' (13). This is the first in a sequence of three *tracking moves* by Sue, which are *supportive* in that they postpone the issue of *compliance or non-compliance*.

Turn 14

Kanzi *responds* with an *affirmative reply* (14). His vocalization appears to be a falling tone, and he points to BEDROOM.

Turn 15

Sue again requests *confirmation* with rising intonation on 'the bedroom' (15).

Turn 16

Kanzi again *responds affirmatively* with a vocalization (16), nodding his head up and down.

Turn 17

Sue makes a third request for *confirmation*, 'you want to Open Bedroom' (17).

Turn 18

Kanzi makes a third *affirmative response*, BEDROOM (18), using the lexical item on the board to provide polar information, i.e. yes it is the bedroom.

Turn 19

Being now quite clear about Kanzi wanting to open the bedroom, Sue makes a *developing move*, 'well Kanzi There Are Toys In The Bedroom For You' (19/a). But as was the case with 'that would be a fun thing to do' (3/b), the high degree of alignment with Kanzi's proposal is more apparent than real, since in her next *move*, Sue doesn't say, for example, 'so let's go play with them in there right now'. What she does say is an *extension* of (19/a), 'We Can Play with those Toys in the Middletestroom' (19/b). Sue's next *move*, 'Kanzi Is staying In Middletestroom Today' (19/c), is unambiguously a third *challenging rejoinder*. In the next three *moves*, Sue *prolongs* her *counter-proposal*, first with an *enhancement*, 'because mr. ida is Grabbing your Picture with his Camera' (19/d), and then with the *elaborations* 'uh huh that's why we're staying in here' (19/e) and 'mm hmm we can do things in here' (19/f). At the same time, Sue's real reason *for non-compliance* (19/d) reveals that the 'reasons' given for her previous *challenging rejoinders*, 'because mike is working on something' (5/b), and 'because we have to go through the grouproom' (9/c) were factitious.

Turn 20

Kanzi hasn't given up negotiating his proposal, however, as he gestures toward the bedroom with his left hand.

Exchange 2 (moves 1 and 2)

Up to this point in the discourse, Sue has been on the defensive. In one way or another, despite the greater number of *moves* she has made, Sue has been dependent on Kanzi; that is, she has had to negotiate the proposal of his *opening move*. At this point, she decides to take the initiative, and begin a new *exchange* by making an *opening move*, perhaps to pre-empt yet another *challenging rejoinder* by Kanzi.

Turn 1

In saying 'what do you want me to get' (1), Sue is making an *offer*, which of course is contingent on Kanzi's provision of lexical information.

Turn 2

By pointing to SURPRISE (2), i.e. by providing the information that is necessary for Sue to carry out her *offer*, Kanzi accepts it.

4.2 Lexicogrammatical patterns

4.2.1 Theory: systematic repetition and cohesive harmony

The analysis in Section 4.1.2 focused on the interaction taking place between Kanzi and Sue through looking at the conversational acts which they produced.

The analysis has not so far investigated the nature of the confrontation through an examination of the lexicogrammatical patterns jointly produced by the two participants. Can an analysis of these lexicogrammatical constructions help gain insight into the nature of the contestation and the way it plays out?

In this endeavor, we would like to take an approach derived from two traditions:

(a) cohesive harmony developed by Hasan (1984), Halliday and Hasan (1985; 1989) and

(b) the notion of systematic repetition developed by Winter (Winter, 1977; 1979; Hoey & Winter, 1981). (See Fries, 1992; 1993 for other analyses that combine cohesive harmony with Winter's notion of systematic repetition within clauses that enter into matching relations).

Both of these approaches examine parallels in lexicogrammatical expression, and therefore depend on a detailed lexicogrammatical analysis.

Let us begin by describing the notion of systematic repetition. It is Winter's contention that when clauses appear in a matching relation such as contrast, alternation or comparison, the clauses must express similar ideas. Winter believes the full expression of contrast goes through four steps.

The four steps are given as follows:

General patterns for expressing contrast

Step 1:	General comparison – focus on difference	Optional
	A differs from B	
Step 2:	Statement of basis of contrast	Obligatory
	X is true of A	

Step 3:   Denial of validity of standard for compared item   Optional
           X is not true of B
Step 4:   Correction of Step 2   Obligatory
           (What is valid for B. This step provides the focus
                of the contrast)
           Y is true of B

As indicated by the term 'optional' beside Steps 1 and 3, not all expressions of contrast go through all these four steps. Rather, expressions of contrast are typically much shorter. The following set of examples illustrates the four steps, and how they may combine ((c) was the original example).

Examples of how the pattern may play out in text

(a) **Complete** (strong emphasis/highlight on the contrast)

   (1) These two volumes differ.

   (2) The first volume describes all the non-passerine birds so far found in Australia.

   (3) The second volume doesn't describe any non-passerine species of birds.

   (4) Rather, it describes the passerine species in Australia.

(b) **Delete denial** (delete Step 3) (less focus on contrast between the volumes)

   (1) These two volumes differ.

   (2) The first volume describes all the non-passerine birds so far found in Australia.

   (4) The second volume describes the passerine species in Australia.

(c) **Delete the general statement** (delete Step 1) (conveys a minimum focus on the difference)

   (2) The first volume describes all the non-passerine birds so far found in Australia.

   (4) The second describes the passerine species.

Winter's model of contrast implies two major points. First, negation plays an important role in the interpretation of contrast. Whether or not Step 3, with its overt expression of negation, is actually present in a text, when we see a relation of contrast between two sentences (Step 2 and Step 4), we also necessarily interpret a negation as being involved. That is, Step 4 implicitly contradicts Step 2. Thus, in our set of examples above, the relation between Sentences 2 and 4 implies a negation whether we take option (a), (b), or (c). That is, we interpret Sentence 4 in each of the three versions as implying that the second volume does **not** describe non-passerine birds.

A second important point is that contrast implies systematic repetition. That is, two clauses which are in contrast with one another do not simply express similar ideas, there must be patterns to the similarities and differences. To have contrast, we must have a background of similarity and a focus of difference. We can rewrite option (c) in the example above in such a way as to highlight this focus of contrast within a framework of partial repetition. In Table 1, we have filled in and marked with brackets information that is understood (e.g. by ellipsis).

	I	II	III	IV
2	The first volume	describes	all the non-passerine birds	so far found in Australia.
4	The second [volume]	describes	the passerine species [of birds]	[so far found in Australia].

Table 1: Alignment of two sentences in contrast in such a way as to highlight the similarities and differences

Columns II and IV contain information that is repeated in the interpretations of the two sentences. Columns I and III contain the focus of difference. However, even in the columns in which there is a focus of difference, there must be a background of similarity and a focus of difference. In Column I, the background of similarity lies in that in both Sentence 2 and Sentence 4, the member of Column I refers to some volume of this book. The focus of difference lies in that these portions of the two sentences refer to *different* volumes. Similarly the background of similarity in Column III lies in that both sentences refer to various bird species. The focus of difference is that the two sentences refer to *different* sets of species. This sort of alignment of sentences in contrast helps the analyst determine what is assumed to be the common background, and what is assumed to be the focus of contrast in these sentences.

Clearly an examination of systematic repetition can benefit from a careful examination of exactly what is repeated and what is in contrast. One technique to discover repeated information in the messages of a text is to look at the cohesive ties, and the similarities and differences in the messages involved. Cohesive harmony, developed by Ruqaiya Hasan, addresses exactly these issues.

Cohesive harmony (Hasan, 1984; Halliday & Hasan, 1985) works on the assumption that most extended coherent texts take some topic and develop it in some way. An adequate analysis of such a text should capture such topic-centered development. Thinking intuitively, a topic-centered text should have relatively many messages which involve the same (small set of) participant(s) – the topic. Further, the messages of the topic-centered text should generally revolve around a small set of ideas.

How can one capture such topic-centeredness? As a first step one can examine the various chains of coreferential items in the text. Coherent texts regularly contain a limited set of referents which recur throughout the text. Each chain of references to an entity or small group of entities can be called an identity chain. The number of the identity chains in a text will provide a first measure of how many referents play a prominent role in the text. (Does the text center around a single main topic – i.e. one identity chain – or do several topics enter into the text?)

As a second level of analysis we can examine the words used in the text, noting how words of similar meaning are distributed in chains through-out the text. Words which are related by any of the semantic relations of antonymy, synonymy, hyponymy, meronymy or strong collocational association are considered to express similar meanings. Chains of words which express similar meanings are called 'similarity chains'. Finally we can look at the similarities among the messages of the text. This is done by examining interactions among the various chains in the text. Two chains interact if and only if the same experiential relation holds between them in at least two clauses. The analysis of chain interaction will provide a measure of how much repetition of similar ideas the text provides.

4.2.2   Second analysis: lexicogrammatical patterns in GO OPEN GROUPROOM

Appendix 2 provides a partial analysis of the 'go-open grouproom' text focusing on the interactions which involve Sue and Kanzi as Actors, material processes, and Locations or Goals. It also addresses a few other messages which are closely related to that set of interactions.

In Appendix 2, the first column provides the move/turn/clause (as in Appendix 1). The second column provides the exact text as in Appendix 1. The third column notes every instance in which Sue or Kanzi (or both) are mentioned as Actors of some material process[1]. The fourth and fifth columns deal with polarity or modalities assigned to the material processes in which Sue and Kanzi are Actors of material processes. The sixth and seventh columns are two different types of material processes. Processes which are included in the 'Go' column (Column 6) are middle processes (See Matthiessen, 1995). Middle processes have only one participant, an Actor. Processes which are placed in the 'Open' column (Column 7) are effective processes. These processes involve two participants, an Actor and a Goal. Columns 8 and 9 indicate Circumstances of Location or Goals, respectively for the processes in Columns 6 and 7. Finally, words in Column 10 indicate a temporal circumstance for the material processes in Columns 6 and 7, while the last three columns indicate words which take part in attributive relational processes. Column 11 indicates a Carrier role while Column 13 indicates an Attributive role.

Each row on the Table indicates a message (generally a clause) of some sort. Items are placed in the same column for two reasons. First, these items must be seen as either belonging to the same identity chain or to the same similarity chain. For example, all words placed in Column 3 refer either to Sue or to Kanzi or to both, and therefore these words constitute an identity chain. Similarly, all words placed in Column 8 refer to alternative parts of the LRC available to Kanzi to go to. They are thus related through the semantic relation of meronymy.

A second criterion for placing words in a column is that the words must be related to other chains in this portion of the text in similar ways. Thus, Sue and Kanzi are Actors of the various actions mentioned in Columns 6 and 7. The polarity items in Column 4 and the modal items in Column 5 all apply to the processes in Columns 6 and 7. Column 8, the locations, are all locations for going and playing (the processes mentioned in Columns 6), and the Goals (Column 9) are all Goals for the processes mentioned in Column 7. This constant relation to the remainder of the message is important to placement on the Table. Thus, a number of references to the grouproom (in S1/3/d – S1/5/b) are not placed in the 'Location' column because these instances fill the Carrier participant role, not the Location participant role in their clause.[2] In other words, although an effort to maintain the original word order has been made, placement in columns left to right represents common participant roles of similar items in similar messages, not word order.

A further point should be made here. At several places, information has been filled in by interpretation. Thus, (K1/1) 'GO OPEN GROUPROOM' is addressed to Sue. The normal way to interpret a command is to take the addressee as the Actor of any action that is requested. Therefore, even though this sentence does not mention Sue, her name has been placed in the appropriate column (in square brackets to indicate that this word did not actually occur in that clause) to show that Sue is the implied Actor of these processes.

Similarly, in (S1/3/b), Sue says 'that would be a fun thing to do.' The word *that* tells the listener that he/she knows what is being referred to and to go find the referent. As listeners, we have interpreted this word to refer to the proposal under negotiation: [Sue go open group room]. To indicate the relevance of this interpretation, the symbol '–' has been placed in each of the relevant columns of that message to indicate the meanings we have interpreted.[3]

The word *grouproom* has been entered twice in Clauses (K1/1), (S1/1), and (S1/3/b). This device was used to show that *grouproom* plays two simultaneous roles in these clauses. On the one hand it functions as a Circumstance of Location for the process of *go*. In addition, however, it also serves as the Goal for the process *open*.[4]

Clearly, the text is organized around a struggle in which one participant proposes alternative places for the activity that is about to take place. Can this analysis detect the fact that these destinations are being proposed as **alternatives**?

The first five utterances essentially repeat one another with little change. The first case where there is significant change is in (S1/3/c) in which Sue adds the modality *wants to*. As pointed out in Section 4.1.2, this modality takes the process out of the here/now/real and makes it unreal. Clause (S1/3/c) is followed by a clause which is introduced with a *but*.

This *but* marks a concessive relation between the two clauses. That is, Clause (S1/3/c) sets up a chain of expectations. The *but* in Clause (S1/3/d) indicates that the expectation will not be fulfilled. One might make that contradiction explicit by filling in an intermediate step. But notice that even with the intermediate step expressed, the paradigm is not complete since Sue does not actually express the final step in her argument. In order to make her argument complete, we need to supply the unspoken conclusion. In order to present the logical and matching relations involved in Sue's concession, we have added the implicit material and marked it as implicit by placing it between square brackets.

a.   Sue wants to open group room

b.   [so   we're probably going to open the group room]

c.   but   the grouproom is broken

d.   [so   we can't/won't open the grouproom]

The contrast implied in the original is made explicit in the systematic repetition between (b) and (d).

It is clear from Kanzi's reaction that he understands the implication in Sue's contribution that they will not go to the grouproom, for immediately after Sue finishes telling him that the grouproom is broken, he says *playyard*. Sue interprets this as suggesting an alternative place to go. (That is, she interprets it as fitting into Column 8 within the set of interactions described by Columns 3 to 10.)

At first glance, Sue's first substantive response (S1/9/a) looks like she is supporting his proposal. It begins with *yes* and uses positive polarity. Like Kanzi, Sue omits to mention the Actors, the process and polarity or modality. However, included in her clause is the adverb *tomorrow*. This adverb clearly presupposes and modifies the (implied) process and produces a configuration that is clearly not what Kanzi wants. Indeed, Sue immediately follows her apparently supportive contribution with a direct denial of what she understands Kanzi is requesting (playyard today).

Winter (1977: 165) found 'With all denials, the most predictable clause relations are either reasons for the denial or corrections of the denial.' (See also Winter, 1977: 490.) Sue has told Kanzi what is not going to happen; once she has finished explaining why Kanzi can't go to the playyard, her next move (S1/9/e) is to tell Kanzi what **will** happen.

The simple statement with positive polarity associated with *will* (a type of future tense here) and *today* associates the assertion with the here and now and real. Sue's succeeding assertions (S1/11/f, S1/11/a–b) all emphasize that point. Indeed (S1/11/b) (*That is what we're going to do*) is interesting in how internally repetitive (and therefore emphatic) it is. Structurally it is a thematic equative clause (or pseudo-cleft, see Halliday, 1994: 42) in which the two parts *that* and *what we're going to do* are nominals[5] which are identified and equated. The word *that* refers to the proposition that has just been presented previously – *we will play here today*. But in addition the word *do* in the sentence is a substitute for the Residue *play here today*. Thus the second part of this clause is also equivalent to 'we're going to play here today'.

After Sue has been so emphatic in her denial of his proposition, Kanzi still tries one more time with the lexigram *bedroom*. Again, after a certain amount of checking on what Kanzi means, Sue interprets Kanzi's *bedroom* as fitting into the message paradigm with *bedroom* naming an alternative place to open[6]. Kanzi confirms this interpretation by reasserting *bedroom*. Sue responds with a statement that displays systematic repetition with her interpretation of Kanzi's request and thus implies a denial of that request and a correction saying what they **will** do. This positive affirmation recurs three times in this portion of the conversation (S1/19/b), (S1/19/c), and (S1/19/e).

The final utterance in this exchange is Kanzi pointing to the bedroom. Interpreting this utterance is difficult. It is clear that Sue interpreted it as a request to get something **from** the bedroom. In this case this utterance would seem to initiate the next exchange, and more to the point for this analysis, would **not** be analyzed as fitting into the paradigm 'go open grouproom' that has been so prominent in the first exchange. On the other hand, this form is compatible with a simple reassertion of the request, 'go open bedroom'. In this case, the utterance would simply extend the message paradigm of the first exchange one more turn.

From the analysis above, it is clear that Kanzi's contributions are made and interpreted as relevant to the ongoing dynamics of the flow of conversation. It is possible to argue that Kanzi's contributions after (K1/1) all involve some sort of grammatical ellipsis. It is not necessary to argue this position, however, to interpret them. His contributions are all relevant to the patterns of meanings created and exhibited through the communications of the two participants. Notice that many of Sue's contributions are also partial clauses. In these cases, since we believe Sue knows English grammar, we have the grammatical device of ellipsis to fall back on to interpret the sentence internal grammar. However, even with her contributions, we interpret their relevance by reference to the patterns of meaning created through the various turns of the participants. Thus, even when she does not use elliptical constructions, she may leave her logic implicit (as in clauses S1/3/c&d). Kanzi is fully capable of filling in the implicit information and interpreting the thrust of Sue's contribution. Even without looking at the grammatical ellipsis, with all such an analysis implies for the control of grammar, this conversation is far more than simply 'handfuls of words strung together'.

5   Conclusion

We now return to our three questions.

1)   What kind of social world are Kanzi and Sue jointly construing?

Despite the limitations of the lexigram board, and minimal, but not non-existent syntax (see Greenfield and Savage-Rumbaugh, 1990), Kanzi's words are distributed systematically according to the rules of casual conversation. A social world *is* being jointly construed in this discourse. It is emphatically *not* 'just handfuls of words strung together'.

First, it is a social world organized by jointly construed discourse which plays out the 'tension between... establishing solidarity through the confirmation of similarities, and... asserting autonomy through the exploration of differences' (Eggins & Slade, 1997: 22; cf Tomasello, 1999: 170 cited above). Sue is doing her best to create an alignment with Kanzi, such that he will abandon his original proposal. Kanzi is doing his best to assert his autonomy by coming up with acceptable alternatives to his original proposal.

Second, it is a social world organized by jointly construed, sequenced complexity. Kanzi deals with Sue's mixed messages of confrontation and support with some considerable skill. At no point does Kanzi interrupt Sue (or Sue Kanzi for that matter). He waits until Sue's three quite different sequences of challenging moves (3/a–d, 5a/b; 9/a, 11a/b; 19/a–f) are 'complete', before he reacts, although in the case of the third sequence, of course, his reaction is pre-empted by Sue's new opening move. Moreover, *both* Sue *and* Kanzi provide partial information in their turns, fully expecting that that partial information will be interpreted as relevant to and in terms of the ongoing interaction. Further, they both are successful in interpreting what is said by the other.

Third, and most important, it is a social world organized by turn-taking. Consideration of discourse as turn-taking provides a particularly strong bit of evidence in support of the contention that Kanzi's meaning-making abilities have been underestimated. In Exchange 1 there are 20 turns and 35 moves. Excluding the initial co-turn, Kanzi and Sue take an equal number of turns (10 vs. 9), although Sue makes over twice as many moves as Kanzi (24 vs. 10). The point is not that Sue makes more moves per turn than Kanzi, but that turn-taking provides the ideal environment for 'role-reversal imitation', the capstone stage of a process of cognitive development that begins with a child's understanding of others as intentional agents (Tomasello, 1999: 103–7).

Tomasello (1999: 105) describes 'role-reversal imitation' this way:

> the child must learn to use a symbol toward the adult in the same
> way the adult used it toward her. This is clearly a process of imitative
> learning in which the child aligns herself with the adult in terms of
> both the goal and the means for attaining that goal; it is just that in
> this case the child must not only substitute herself for the adult as
> actor (which occurs in all types of cultural learning) but also substitute
> the adult for herself as the target of the intentional act (i.e. she must
> substitute the adult's attentional state as goal for her own attentional
> state as goal).

Kanzi deploys speech function symbols in precisely this way. Kanzi's registers and challenges target Sue in the same way that Sue's registers and challenges target Kanzi.

Tomasello (1999: 106) continues:

> the result of this process of role-reversal imitation is a linguistic
> symbol: a communicative device understood intersubjectively from
> both sides of the interaction. That is to say, this learning process
> ensures that the child understands that she has acquired a symbol
> that is socially 'shared' in the sense that she can assume in most
> circumstances that the listener also comprehends and can produce that
> same symbol and the listener also knows they can both comprehend
> and produce the symbol. The process of understanding communicative

signals – as in chimpanzee and some prelinguistic infant gestural communication – is very different in that each participant understands its role only, from its own internal perspective.

Perhaps the clearest example of 'role-reversal imitation', and thus of a canonical linguistic symbol is Kanzi's original command. Savage-Rumbaugh et al. (1998) gives an account of one experiment in which Kanzi was given 660 novel commands, e.g. 'Put the toothbrush in the lemonade'. Kanzi carried out 72 per cent of these commands successfully, and clearly understood his role to be compliance. In the opening move of the first exchange 'GO OPEN GROUPROOM', their roles are just as clearly reversed: Sue is expected to be in compliance, and when she is not, she remains targeted, with alternative routes to compliance.

2)   To what extent does Kanzi participate in the human, continuous negotiation of differential power relations?

The particular difference that is being negotiated between Kanzi and Sue in this discourse is power. The discourse enacts a power struggle. First, Kanzi has taken on the role of opening and initiating a discourse. This is very different from the role of sustaining. It is a claim to control. And a command is a powerful speech function. When Kanzi says 'GO OPEN GROUPROOM', the onus of compliance rests just as squarely on Sue's shoulders as it would rest on Kanzi's, if Sue were to say to Kanzi, for example, 'Go to the office and bring back the red ball'. Second, by making an opening move, Kanzi sets the agenda for negotiation. This makes Sue grammatically dependent on him, since she has to negotiate *his* proposal. In exchange 2, however, Kanzi's acceptance of Sue's offer is a tacit admission that he has failed in his challenge to her power: they will stay in the middle test room. The point here is not just that Sue will get toys from a room to which she, but not Kanzi, has ready access, but that this asymmetry of power has been enacted in their joint language (including gestural and vocal realizations of moves on Kanzi's part).

3)   To what extent might discourse be the motivational environment for the development of both interpersonal grammar (mood) and ideational grammar (processes, participants, and circumstances)?

Kanzi's facility in enacting the symbolic roles of casual conversation suggests that discourse is an environment favourable to symbolic exchange of goods and services and also of information. Thus discourse provides a symbolic environment favourable to the development of symbolic representation of transitivity roles and functions. (For a complementary view which derives syntax from discourse, see Rolfe, 1999: 776–92.)

The dialogue analysed in this paper revolves around the exchange of goods and services. Thus, both Kanzi's opening move 'GO OPEN GROUP ROOM', and Sue's opening move 'What Do You Want Me To Get' foreground this issue. At the same time, Kanzi's command and Sue's offer require the expression of processes and participants independent of the interpersonal role relationships. The packaging of Kanzi's command conceals what could be otherwise construed as involving two processes, whereas the packaging of Sue's offer reveals these two processes.

In the grammar of the first exchange, Kanzi's command 'GO OPEN GROUP ROOM' expresses a material process of 'doing' and a Range (Halliday, 1994: 146–9) for the doing, and requires an Actor (the addressee Sue). What is concealed by the 'doing' is the mental process of 'wanting'. Because of their differential power relation, Kanzi assigns Sue a material 'doing', but implicitly Kanzi assigns himself both a mental 'wanting' and a Beneficiary role, i.e. Sue's 'doing'–compliance both satisfies his desire, and is done for his benefit. In other words, a semantic space is opened up for expressing two processes and two participants, as for example in the command Kanzi didn't use, 'WANT SUE OPEN GROUPROOM' (Halliday, 1994: 289). Here there is a Senser (Kanzi) and a mental process ('WANT'), an Actor ('SUE'), a material process ('OPEN'), and a Range ('GROUPROOM').

In the grammar of the second exchange, Sue's offer 'What Do You Want Me To Get' reveals rather than conceals the 'wanting'/'doing' distinction, and by foregrounding the complementarity of offers and commands, reinforces the motivation for the expression of two processes and two participants. An offer is the mirror image of a command. It pre-empts a command by anticipating it, and thus begins with compliance, for example, 'I'll get you a surprise, shall I?', where the Beneficiary role ('you') is made grammatically explicit. But there is an additional complexity in Sue's offer, because her 'what' makes the exchange of goods and services contingent on an exchange of information, in this case 'a surprise', which constitutes acceptance of the offer.

In short, the kind of complexity involved in the opening moves (offer, command, statement, and question) and responses in discourse creates an enormous motivation to develop the grammatical tools to accomplish these discourse moves. These grammatical tools include *both* a symbolic interpersonal lexicogrammar, with contrasts among imperative, declarative, interrogative moods, *and* a symbolic ideational lexicogrammar, with participants, processes, and circumstances, in order to have something to negotiate about, and to map the interpersonal and ideational grammars onto each other.

Notes

Peter Fries's contributions to this paper were written while on sabbatical from Central Michigan University. This article is a revised version of a paper given at the Second International Workshop of the Systemic Functional Research Community on Interpersonal and Ideational Grammar (University of Leuven, 21–24 November 2001). The Research Community is funded by the Fund for Scientific Research – Flanders (Belgium) (grant no. WO.018.00N).

1   See Halliday (1994) for a description of the experiential metafunction of language: the process types, the participant roles and the Circumstances that are involved.

2   These occurrences of *grouproom* **are**, however, members of the identity chain which refers to the grouproom in this text.

3   Note that the entire clause 'That would be a fun thing to do' is a relational clause and so looks similar to *grouproom is broken*. This clause (S1/3/b) does not belong in the same paradigm as *grouproom is broken*, however, since 'fun thing to do' is an attribute which has no lexical similarity to *broken*, and the Carrier (*that*, which stands for the entire proposal) also does not belong to any similarity or identity chain which is involved in interactions in any of the other clauses in Appendix 2.

4   Of course the analysis adopted here implies that we analyze *go* and *open* as two unrelated verbs. This analysis is not automatic. We might have analyzed *go* as a kind of (motional) modality on *open* and thus not a separate process in itself. In which case, *go open* would have been analyzed as a single process and the dual interpretation of *grouproom* would not have been relevant. The treatment used in this paper is similar to the discussion by Greenfield and Savage-Rumbaugh (1990: 564–6) for conjoined actions.

5   *That* is obviously a nominal in that it is regularly analyzed as a pronoun. The other part of the clause which is equated to *that* is an embedded clause – a clause that has become nominal in nature.

6   But also note the close association of *go* and *open*. Thus *bedroom* here
    is not only a Goal for *open* but also a Location for an implied *go*. A full
    description of the process types is beyond the scope of this paper. The
    concepts used are described in Butt et al. (2000), Halliday (1994) and Mat-
    thiessen (1995), among other sources.

References

Butt, D., Fahey, R., Feez, S., Spinks, S. and Yallop, C. (2000) *Using Functional
    Grammar: an explorer's guide*. Sydney, Australia: National Centre for
    English Language Teaching and Research, Macquarie University.

Calvin, W. H. and Bickerton, D. (2000) *Lingua ex Machina: reconciling Darwin
    and Chomsky with the human brain*. Cambridge, MA: MIT Press.

Deacon, T. W. (1997) *The Symbolic Species: the co-evolution of language and
    the brain*. London: W. W. Norton.

Eggins, S. and Slade, D. (1997) *Analysing Casual Conversation*. London:
    Cassell.

Fries, P. H. (1992) Lexico-grammatical patterns and the interpretation of texts.
    Discourse Processes 15: 73–91.

Fries, P. H. (1993) On repetition and interpretation. In S. K. Verma and V.
    Prakasam (eds) *New Horizons in Functional Linguistics*. pp. 69–102.
    Hyderabad, India: Booklinks.

Greenfield, P. M. and Savage-Rumbaugh, S. (1990) Grammatical combination
    in *Pan paniscus*: processes of learning and invention in the evolution and
    development of language. In S. Parker and K. Gibson (eds) *'Language' and
    Intelligence in Monkeys and Apes: comparative developmental psychology
    of language and intelligence in primates*. Cambridge: Cambridge University
    Press.

Halliday, M. A. K. (1975) *Learning How to Mean: explorations in the develop-
    ment of language*. London: Edward Arnold; New York: American Elsevier.

Halliday, M. A. K. (1994) *An Introduction to Functional Grammar*. London:
    Edward Arnold.

Halliday, M. A. K. and Hasan, R. (1985) *Language, Context, and Text: aspects
    of language in a social-semiotic perspective*. London: Oxford University
    Press.

Hasan, R. (1984) Coherence and cohesive harmony. In J. Flood (ed.)
    *Understanding Reading Comprehension: cognition, language and the struc-
    ture of prose*. pp. 181–219. Newark, DE: International Reading Association.

Hoey, M. and Winter, E. (1981) Believe me for mine honour. *Language and
    Style* 14: 315–39.

Matthiessen, C. M. I. M. (1995) *Lexicogrammatical Cartography*. Tokyo:
    International Language Sciences Publishers.

Matthiessen, C. M. I. M. (2004) The evolution of language: a systemic functional exploration of phylogenetic phases. In G. Williams and A. Lukin (eds) *The Development of Language*: *functional perspectives on species and individuals*. London: Continuum.

Painter, C. (1984) *Into the Mother Tongue*: *a case study in early language development*. London: Pinter.

Painter, C. (1999) *Learning Through Language in Early Childhood*. London: Cassell.

Pinker, S. (1994) *The Language Instinct*. New York: William Morrow.

Rolfe, L. (1999) Theoretical stages in the prehistory of grammar. In A. Lock and C. R. Peters (eds) *Handbook of Human Symbolic Evolution*. Oxford: Blackwell.

Savage-Rumbaugh, S., Shanker, S. and Taylor, T. (1998) *Apes, Language, and the Human Mind*. New York: Oxford University Press.

Savage-Rumbaugh, S. and Lewin, R. (1994) *Kanzi*: *the ape at the brink of the human mind*. New York: John Wiley.

Tomasello, M. (1999) *The Cultural Origins of Human Cognition*. Cambridge, MA: Harvard University Press.

Winter, E. O. (1977) *Replacement as a Function of Repetition*: *a study of some of its principle features in the clause relations of contemporary English*. PhD Dissertation, University of London. University Microfilms #77–70,036.

Winter, E. O. (1979) Replacement as a fundamental function of the sentence in context. *Forum Linguisticum*. 4: 95–133.

Appendix 1

Speech function codings for GO OPEN GROUPROOM

functions	exchange	turn/move	speaker	text
open: initiate: command	1	1 1 [this is a co-turn]	Kanzi Sue	(i) GO OPEN GROUPROOM (i) go open grouproom
sustain: continue: prolong: enhance	1	2	Kanzi	gesture toward the grouproom (assumed)
sustain: react: respond: support: register	1	3/a	Sue	(i) oh over there
sustain: react: respond: support: develop: elaborate	1	3/b	Sue	(i) that's that would be a fun thing to do
sustain: continue: prolong: elaborate	1	3/c	Sue	(i) Yes Sue Wants To Open Grouproom
sustain: react: rejoinder: confront: challenge: rebound	1	3/d	Sue	(i) but Grouproom Is Broken

functions	exchange	turn/ move	speaker	text
sustain: react: respond: support: register	1	4	Kanzi	(i) BROKEN (Kanzi looks away not questioningly at Sue)
sustain: react: rejoinder: support: response: resolve	1	5/a	Sue	(i) yeah it's broken (ii) it's broken
sustain: continue: prolong: enhance	1	5/b	Sue	(i) it's broken in there (ii) because mike is working on something
sustain: react: rejoinder: confront: challenge: counter	1	6	Kanzi	(i) PLAYYARD
sustain: react: rejoinder: support: track: confirm	1	7	Sue	(i) playyard
sustain:	1	8	Kanzi	vocalizing
sustain: react: respond: support: register	1	9/a	Sue	(i) oh

functions	exchange	turn/ move	speaker	text
sustain: react: rejoinder: confront: challenge: counter	1	9/b	Sue	(i) well Yes Playyard… Tomorrow
sustain: continue: prolong: enhance	1	9/c	Sue	(i) we can't go to the playyard today (ii) because we have to go through the grouproom
sustain: continue: prolong: extend	1	9/d	Sue	(i) and the Grouproom Is Broken
sustain: continue: prolong: enhance	1	9/e	Sue	(i) Today We Will Play Here
sustain: continue: prolong: elaborate	1	9/f	Sue	(i) In Middletestroom
sustain: react: respond: support: register	1	10	Kanzi	points towards MIDDLETESTROOM lexigram
sustain: react: respond: support: reply: affirm	1	11/a	Sue	(i) yeah yes yes middletestroom mm hmm

functions	exchange	turn/ move	speaker	text
sustain: continue: prolong: elaborate	1	11/b	Sue	(i) mm hmm that's what we're going to do
sustain: react: rejoinder: confront: challenge: counter	1	12	Kanzi	(i) BEDROOM
sustain: react: rejoinder: support: track: confirm	1	13	Sue	(i) and you're pointing (ii) are you pointing to the Bedroom
sustain: react: respond: support: reply: affirm	1	14	Kanzi	vocalization/BEDROOM [vocalization appears to be falling tone]
sustain: react: rejoinder: support: track: confirm	1	15	Sue	the bedroom
sustain: react: respond: support: reply: affirm	1	16	Kanzi	vocalization (nods head up and down)

functions	exchange	turn/move	speaker	text
sustain: react: rejoinder: support: track: confirm	1	17	Sue	(i) you want to Open Bedroom
sustain: react: respond: support: reply: affirm	1	18	Kanzi	(i) BEDROOM
sustain: react: respond: support: develop: extend	1	19/a	Sue	(i) well kanzi There Are Toys In The Bedroom For You
sustain: continue: prolong: extend	1	19/b	Sue	(i) We Can Play with those Toys in the Middletestroom [the coding choice here lay between this 'supportive' interpretation and calling Sue's move a counter]
sustain: react: rejoinder: confront: challenge: counter	1	19/c	Sue	(i) Kanzi Is staying In Middletestroom Today
sustain: continue: prolong: enhance	1	19/d	Sue	(i) because mr. ida is Grabbing your Picture with his Camera [if 19/b is interpreted as counter, then 19/c is elaborate]

functions	exchange	turn/ move	speaker	text
sustain: continue: prolong: elaborate	1	19/e	Sue	(i) uh huh that's why we're staying in here
sustain: continue: prolong: elaborate	1	19/f	Sue	(i) mm hmm we can do things in here
sustain:	1	20	Kanzi	Kanzi gestures towards the bedroom with his left hand
open: initiate: offer	2	1	Sue	(i) What Do You Want Me To Get
sustain: react: respond: support: reply: accept	2	2	Kanzi	(i) SURPRISE

Appendix 2

Systematic repetition in GO OPEN GROUPROOM

1	2	3	4	5	6	7	8	9	10	11	12	13
		Actor										
		Sue / Kanzi	Pol	modal	go	open	Location	Goal	Temp	Carrier	Rel	Attr
(K1/1)	GO OPEN GROUPROOM	[Sue]	pos		go	open	grouproom	group room				
(S1/1)	go open grouproom	[Sue]	pos		go	open	grouproom	group room				
(K1/2)	[gestures toward the grouproom]		pos				grouproom					
(S1/3/a)	oh over there that's that would be a fun thing		pos				there					
(S1/3/b)	[[to do]]	'_'			'_'	'_'	'_'	'_'				
(S1/3/c)	Yes Sue Wants To Open Grouproom	Sue	pos	to	wants	open		group room				
(S1/3/d)	but Grouproom Is Broken									group room	is	broken
(K1/4)	BROKEN											broken
(S1/5/a.i)	yeah it's broken									it	s	broken
(S1/5/a.ii)	it's broken									it	s	broken
(S1/5/b)	it's broken in there because mike is working on									it	s	broken
(S1/5/b)	something											
(K1/6)	PLAYYARD		pos				playyard					
(S1/7)	playyard		pos									
(K1/8)	[vocalizing]		pos									
(S1/9a)	oh well Yes Playyard...											
(S1/9/b)	Tomorrow		pos				playyard		tomorrow			
(S1/9/c.i)	we can't go to the playyard today	we	neg	can	go		to the playyard		today			

1	2	3	4	5	6	7	8	9	10	11	12	13
		Actor										
		Sue / Kanzi	Pol	modal	go	open	Location	Goal	Temp	Carrier	Rel	Attr
	because we have						through					
	to go through the			have			the					
(S1/9/c.ii)	grouproom	we		to	go		grouproom					
	and the											
	grouproom is									group		
(S1/9/d)	broken									room	is	broken
	Today We Will											
(S1/9/e)	Play Here	we		will	play		here		today			
							in					
							middletest					
(S1/9/f)	In Middletestroom						room					
	[points toward											
	MIDDLE											
(K1/10)	TESTROOM]						[here]					
	yeah yes yes											
	middletetsroom						middletest					
(S1/11/a)	mm hmmm		pos				room					
	mm hmm that's											
	[[what we're											
(S1/11/b)	going to do]]	'–'		'–'	'–'	'–'	'–'					
				re								
	[[what we're				going							
(S1/11/b.i)	going to do]]	we	pos	to	do [[?]]							
(K1/12)	BEDROOM		pos				BEDROOM					
	and you're											
(S1/13.i)	pointing											
	are you pointing											
(S1/13.2)	to the Bedroom											
	vocalization/											
(K1/14)	BEDROOM		pos				BEDROOM					
							the					
(S1/15)	the bedroom		pos				bedroom					
	[vocalization											
	– nods head up											
(K1/16)	and down]		pos									

1	2	3	4	5	6	7	8	9	10	11	12	13
		Actor										
		Sue / Kanzi	Pol	modal	go	open	Location	Goal	Temp	Carrier	Rel	Attr
(S1/17)	you want to Open Bedroom [BEDROOM – K uses lexical info to provide yes no	[Kanzi]	pos	want to		open	the bedroom					
(K1/18)	info]		pos				BEDROOM					
(S1/19/a)	well kanzi There Are Toys In The Bedroom For You											
(S1/19/b)	We Can Play with those Toys in the Middletestroom	we	pos	can	play		in the middle testroom					
(S1/19/c)	Kanzi Is Staying In Middletestroom Today	Kanzi	pos		is staying		in the middle testroom		today			
(S1/19/d)	because mr. ida is Grabbing your Picture with his Camera											
(S1/19/e)	uh huh that's why we're staying in here	we	pos		are staying		in here					
(S1/19/f)	mm hmm we can do things in here [Kanzi gestures toward the bedroom with his											
(K1/20)	left hand]						bedroom					
(S2/1)	What Do You Want Me To Get											
(K2/2)	SURPRISE											

2 The ideational dimension: evidence for symbolic language processing in a bonobo (Pan paniscus)

James D. Benson, William S. Greaves,
Michael O'Donnell and Jared P. Taglialatela

Abstract

Evidence that an animal is capable of some degree of symbolic, human language processing supports the argument that the animal's consciousness is to some degree human-like. In this chapter, we reinterpret the findings of Savage-Rumbaugh et al. (1993) using the twin tools of Deacon's referential hierarchy and Systemic Functional Linguistics, with a view to providing further corroborative evidence for a bonobo ape's symbolic processing abilities, and as a result to open a window into the consciousness of at least one non-human primate.

1 Introduction: consciousness and language

Recent neurobiological proposals differentiate layers of consciousness. Edelman and Tononi's (2000) 'higher-order consciousness' depends on 'primary consciousness'. Language is the domain of the former, but not the latter. Similarly, Damasio's (1999) 'extended consciousness' depends on 'core consciousness', and language is the domain of the former, but not the latter. These proposals, however, are not identical (Damasio, 1999: 338), since

> [Edelman's] primary consciousness is simpler than my core consciousness and does not result in the emergence of a self. Edelman's higher-order consciousness is also not the same as my extended consciousness, because it requires language and is strictly human.

In either case, if there were evidence that an animal was capable of some degree of symbolic, human language processing, it could be argued that the animal's consciousness was to some degree human-like.

Although evidence for symbolic processing in non-human primates is extensive, as in Fouts and Waters' (2001) review of over 30 years of chimpanzee sign-language studies, our focus here is a narrow one: a small part of one study of one animal, the bonobo Kanzi, as reported in Savage-Rumbaugh et al. (1993), *Language Comprehension in Ape and Child*. Evidence of Kanzi's symbolic processing abilities would provide support for Damasio's (1999: 198) belief that bonobos employ extended consciousness:

> I also believe that apes such as bonobo chimpanzees have an autobiographical self, and I am willing to venture that some dogs of my acquaintance also do.

For Edelman and Tononi (2000: 194), concepts of self, past or future:

> emerged only when semantic capabilities – the ability to express feelings and refer to objects and events by symbolic means – appeared in the course of evolution. Necessarily, higher-order consciousness involves social interactions. When full linguistic capability based on syntax appeared in precursors of homo sapiens, higher-order consciousness flowered, partly as a result of exchanges in a community of speakers. Syntactical and semantic systems provided a new means for symbolic construction and a new type of mediating higher-order consciousness.

At the very least, then, evidence of Kanzi's symbolic processing abilities would provide evidence for a degree of higher-order consciousness.

The experiment reported in Savage-Rumbaugh et al. (1993) was rigorously designed to determine whether or not Kanzi could successfully carry out novel spoken requests, under carefully controlled conditions. Kanzi's ability to do so would be evidence of his comprehension of syntax. The results from this report suggest that, when compared to a human child, Kanzi does in fact understand syntax.

Some remain sceptical of Kanzi's syntactic abilities (for example Calvin and Bickerton, 2000: 23–4, 38–40). However, during the past decade, an exclusive preoccupation with language as de-contextualized, de-semanticised autonomous syntax has abated and scholars in many fields outside of linguistics with an interest in it have been more inclined to think about language in terms of richly contextualized symbolic processing. Deacon (1997) provides a good example of this approach. In Deacon's (1997) interpretation of Peirce's semiotic, there is a referential hierarchy of icon, index, and symbol: indices are correlations of recognizable icons, and symbols are relations among indices. Systemic Functional Linguistics (SFL), with its highly semanticized grammar and contextual orientation, complements Deacon's (1997) proposals about symbolic interpretation. In particular, SFL provides a theoretical foundation for showing how indices can be integrated into symbolic combinatorial systems. With the twin tools of Deacon's referential hierarchy and SFL in hand, we are able to reinterpret the findings of Savage-Rumbaugh et al. (1993), with a view to providing further corroborative evidence for Kanzi's symbolic processing abilities, and as a result to open a window into the consciousness of at least one non-human primate.

2   SFL concepts: a brief sketch

In order to make our case, we need to develop a brief sketch of the theoretical orientation of SFL. Comprehensive accounts can be found in Halliday (1994); Matthiessen (1995); Halliday and Matthiessen (1999) and a number of introductory works, e.g. Eggins (1994); Eggins and Slade (1997); Martin et al. (1997); Thompson (1996). The web site maintained by Mick O'Donnell[1] includes further bibliographical information.

For SFL, language is a resource for making meanings. Meanings are made according to 'principles of semiotic design' (Halliday, 2001): stratification, metafunction, and composition (syntagm and paradigm), shown in Figure 1. These principles provide a concise summary of the basic concepts of SFL.

Stratification

eco-social environment (context)

language:

• discourse-semantics	layers of higher-order semiotic systems, each of which
• lexicogrammar	are organized differently
• phonology	
• phonetics	
• bodily environment	

Metafunction	ideational	differentiation of communicative function into
	interpersonal	ideational, interpersonal, and textual
	textual	
Composition	syntagmatic axis	structural realization of paradigmatic choice
	paradigmatic axis	system networks of choice

Figure 1: Principles of semiotic design: adapted from Halliday (2001)

In the discussion of Figure 1, we focus on those aspects of the theory which are most directly relevant to Kanzi's symbolic processing.

2.1 Stratification

Stratification identifies a set of nested hierarchies, where each level re-organizes in a different way what is organized in the level above. For example, the discourse-semantic SPEECH FUNCTION system with the terms [statement], [question], [offer], [command] is re-organized by the lexicogrammatical MOOD system with the terms [indicative: declarative], [indicative: interrogative], [imperative]. The relation between strata is one of *realization*; for example the SPEECH FUNCTION term [command] is typically realized by the MOOD term [imperative]. In this paper we consider only the lexicogrammatical level.

2.2 Metafunction

The metafunctions account for different things, and employ distinct modes of organization. The ideational metafunction is concerned with the representation of experience, and is realized in constituent structures. The interpersonal metafunction is concerned with the enactment of attitudes and social roles, and is realized by prosodic structures that are dispersed across constituent boundaries. The textual metafunction is concerned with enabling ideational and interpersonal meaning, such that any stretch of language is coherent with itself and with the eco-social environment, and is realized by periodic structures. In this paper we consider only the ideational metafunction.

2.3 Systems and structures

Systems are formalized as a set of choices with an entry condition, as shown in Figure 2.

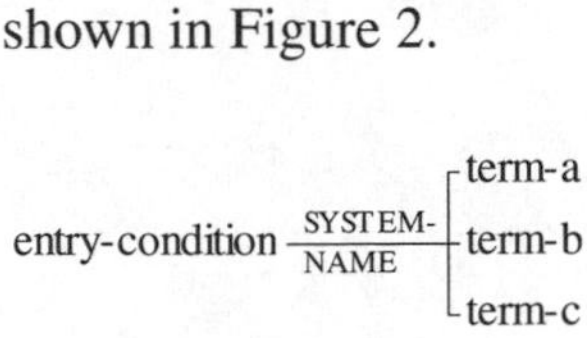

$$\text{entry-condition} \underset{\text{NAME}}{\overset{\text{SYSTEM-}}{\text{---}}} \begin{cases} \text{term-a} \\ \text{term-b} \\ \text{term-c} \end{cases}$$

Figure 2: A system

When two or more systems are related to each other, they form a network, as shown in Figure 3.

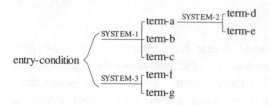

Figure 3: A system network

The relations are of two kinds: delicacy and simultaneity. In delicacy, which is a move in differentiation, a term of the one system becomes the entry condition for the more delicate system. In Figure 3, term-a in SYSTEM-1 is the entry condition for SYSTEM-2.

In simultaneity, there is one entry condition for two or more simultaneous systems. In Figure 3, SYSTEM-1 and SYSTEM-3 share a single entry-condition.

2.4 Paradigmatic and syntagmatic

In SFL, syntagmatic function-structures are derived from paradigmatic systems. Although we are not concerned here with interpersonal systems, we nevertheless illustrate the derivation of function-structures in the interpersonal MOOD system, because it is more generally known than the ideational TRANSITIVITY system. The realization of terms is shown in boxes in Figure 4.

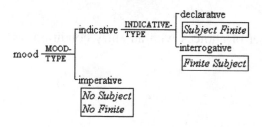

Figure 4: Interpersonal lexicogrammatical system with structural realizations

In this paper we are concerned with ideational TRANSITIVITY systems and structures in general and material processes in particular. At the clause level, AGENCY and PROCESS-TYPE systems yield a structural output of configurations of processes, participants, and circumstances. Such a configuration for a grammarian's example is shown in Figure 5:

Actor	Material Process	Goal	Circumstance: location
Mary	*dropped*	*the letter*	*in the mailbox*

Figure 5: Transitive processing (with participants and circumstance)

In many cases, AGENCY and PROCESS-TYPE systems do not configure with a Goal, but with a Range. A Range, unlike a Goal, cannot be disposed or created, as in the example in Figure 6:

Actor	Material Process	Range	Circumstance: time
they	*climbed*	*the mountain*	*early in the morning*

Figure 6: Transitive processing (with Range instead of Goal)

2.5 Transitive and ergative perspectives

In SFL, there are two perspectives on material processes, the transitive and the ergative. Figure 5 is an example of the transitive perspective, i.e. a 'process and extension model' (Davidse, 1992: 108). Its point of departure is an Actor performing an action, which may extend rightwards to a Goal. The ergative perspective, however, is an 'instigation of process model' (Davidse 1992: 109). In the ergative perspective, the question is not whether the action extends to a Goal, but whether the process is self-instigated through a Medium, or extends leftward to an external Instigator, as illustrated in Figure 7.

	the glass	*broke*
	Medium	Material Process
the cat	*broke*	*the glass*
Instigator	Material Process	Medium

Figure 7: Ergative processing (with participants)

3   The methodology of the experiment reinterpreted in terms of SFL

Language Comprehension in Ape and Child (Savage-Rumbaugh et al., 1993) is a monograph reporting on an experiment which compared the responses of Kanzi, (then, eight years of age), and a two year-old child, Alia, to 660 novel requests. Overall, Kanzi successfully carried out 72 per cent of these requests, compared to 66 per cent for Alia. There were 13 syntactic types and subtypes of requests. We will be looking at a sub-set of 42 paired requests (Savage-Rumbaugh et al., 1993: 95–6), the purpose of which was to examine Kanzi's sensitivity to word order (Savage-Rumbaugh et al., 1993: 92): 'the final group reflects all sentences in which the order of the key words was reversed while maintaining the same verb ('Put the juice in the egg'/ 'Put the egg in the juice')'. As we shall see, however, much more is involved here than word order.

The trials moved from a non-blind situation to a blind one. In order to guard against inadvertently giving cues, the experimenter was stationed behind a one-way mirror, and thus was not visible either to Kanzi or the caretaker who was also present. The first third of the trials were non-blind, because it took some time for Kanzi (and Alia for that matter) to get used to the idea of hearing a disembodied voice making the request. In addition, the caretaker present with Kanzi was hearing loud music through headphones, so as not to hear the request and give anything away. This has important consequences for one type of request in particular, as we discuss later in Section 4.3.2.

3.1 Request types

The paired requests include three of the 13 total request types used in the experiment. What follows is the authors' description of these three types of request, together with our interpretive comments from a systemic perspective.

3.1.1 *Type 1A (Savage-Rumbaugh et al., 1993: 54)*

Put object X in/on transportable object Y (*e.g.*, '*Put the ball on the pine needles*'). These sentences required that the subject construct a relation of adjacency between two objects, *X* and *Y*. Both *X* and *Y* were movable; consequently, actions of this type could be reversed. Objects *X* and *Y* were related by the verb *put* and by a preposition. The prepositions *in* and *on* were used to denote the spatial relations of *X* and *Y*... [T]he goal of the type 1A requests was to determine whether the subjects understood that the sentence required them to construct a relation between X and Y (as opposed to acting independently on X and Y), not to determine whether they fully comprehended the prepositions.

The first type of request locates us squarely within the domain of transitivity: processes, participants, and circumstances. The authors see the preposition as establishing a spatial relationship. This is very close to Halliday's (1994: 212) account of a preposition as a 'minor verb', and of a prepositional phrase as 'a clause-like structure in which the Process/Predicator function is performed by a preposition and not a verb'. Halliday (1994: 213) goes on to say: 'the nominal group corresponds in function to one or other of the participants Range, Goal or Attribute, though without any clear distinction between them. We shall interpret it in all cases as Range'.

Halliday (1994: 213) also says: 'In principal, a non-finite clause implies a potential Subject, whereas a prepositional phrase does not'. But Subject is an interpersonal function. What can we say about ideational functions in the prepositional phrase? Halliday's (1994: 213) example here is a one-participant clause, 'the boy stood on the burning deck', but 'put the ball on the pine needles' is a two-participant-process, with an Actor (implicit, the addressee) and a Goal. In a very real sense, what is spatially related by the preposition ('on'), considered as a Minor Process, are the Goal ('the ball') of the action 'put', and the Range ('pine needles') of the Minor Process. There are two processes here: the action of 'putting' an object, and the end result of the 'putting', a state of rest in which the two objects become related to each other in space in a particular configuration. That is, the Goal ('ball') of the Major Process ('put') becomes the Medium ('ball') of the Minor Process of locating ('on'), which comes to rest at a Range ('the pine needles'). In other words, the transitive perspective shifts to the ergative perspective in mid-clause, as shown in Table 1.

	(Kanzi)	put	the ball	on the pine needles	
transitive perspective	(Actor)	Material Process	Goal	Circumstance: location	
ergative perspective	(Agent)	Major Process	Medium	Minor Process	Range

Table 1: Shift from transitive to ergative perspective

This interpretation more closely matches the authors' understanding of the task, '[T]he goal of the type 1A requests was to determine whether the subjects understood that the sentence required them to construct a relation between X and Y'. The Minor Process captures the 'relation between X and Y' which the authors understand to be the point of this request type.

3.1.2 Type 4 (Savage-Rumbaugh et al., 1993: 60–1)

> *Announce information* (e.g., 'The surprise is hiding in the dishwasher').
> These sentences differ from all the others in that they did not request
> a specific action but simply provided the subject with information.
> Although constructed in a variety of syntactic forms, they are lumped
> together here as a group since their purpose was not to investigate
> syntactic features but to determine whether the subjects would
> respond appropriately if the statement was not in a request format. In
> their everyday lives, most of the sentences heard by Kanzi and Alia
> functioned to provide information rather than to make requests; the latter
> were employed extensively during this test simply because this format
> provides the clearest assay of what it was that the subject understood
> about any given sentence.

> Information that was of interest and designed to promote some response
> was presented to the subjects in order to determine whether they would
> respond to statements as they would to requests. This information was
> limited to two forms: announcements either of where an object was
> hidden or that another party was about to engage in a tickle or chase
> game with them.

The second type of request has an Actor Process Goal configuration, but no Circumstance. Only the second type of announcement figures in the paired requests, i.e. 'that another party was about to engage in a tickle or chase game'. Interestingly, one of these involves a different kind of Circumstance than spatial location, i.e. means: 'Kanzi is going to tickle Liz with the bunny'.

Here we are clearly concerned with metaphorical realizations in lexicogrammar of commands. It is worth noting that the experimental situation of the 660 requests is embedded in discourse, and that this discourse has a generic structure. First the experimenters must get Kanzi's attention, and then they negotiate his state of readiness. Only then can the request be made, and immediately after compliance, Kanzi is praised by the experimenters. This pattern is repeated 660 times, so the discourse moves are unambiguous: the experimenters' role is to command, and Kanzi's role is to comply. In this context, Kanzi, like human agents, can process metaphorical as well as congruent realizations (Halliday, 1994: 342–67), as displayed in Table 2.

SPEECH FUNCTION	Command	
	congruent realization	metaphorical realization
	Tickle Liz with the bunny	*Kanzi is going to tickle Liz* *With the bunny*
MOOD	imperative	indicative

Table 2: Congruent and metaphorical realizations of a command

3.1.3   Type 6 (Savage-Rumbaugh et al., 1993: 66–7)

Make pretend animate A *do action* A *on recipient* Y (e.g., 'Make the doggie bite the snake'). These requests made use of toy objects such as a dog, a snake, a bunny, and so on. The pretend animates were employed only for test purposes, and, as noted earlier, neither subject was specifically taught to transfer from real to pretend exemplars for purposes of this test, although both did so readily. Both subjects understood the application of these different terms to toys as a form of 'pretending' and treated these toys very differently than they would treat the real animal. Additionally, their treatment was appropriate; for example, they patted and hugged the dog but held the snake gingerly.

This third type of request has an Initiatior function (Kanzi), in addition to the functions of Actor ('the doggie'), Process ('make... bite'), and Goal ('the snake') (cf. Halliday, 1994: 286), as shown in Table 3.

(Kanzi)	*make*	*the doggie*	*bite*	*the snake*
(Initiator)	Material...	Actor	...Process	Goal
		Transitive perspective		

Table 3: Material Process with Initiator function

A complete list of the 42 requests is displayed in Table 4. (Adapted from Savage-Rumbaugh et al., 1993: Table 8C, 95–6.) The smaller numbers refer to sequence of paired requests which were extracted from the entire sequence of 660 requests. The larger numbers refer the location of each request in the 660. 'C' indicates a correct response, 'PC' indicates a partially correct response, and 'I' indicates an incorrect response.

Pairs	Running #	Score	Sentences
1	(110)	C	*Can you put some oil on your ball?*
2	(516)	C	*Put the ball in the oil.*
3	(569)	PC	*Put the hat on your ball.*
4	(599)	I	*Put the ball on the hat.*
5	(48)	C	*Put the ball on the rock.*
6	(95)	C	*Can you put the rock on your ball?*
7	(251)	C	*Put the pine needles in your ball.*
8	(588)	C	*Can you put the ball on the pine needles?*
9	(701)	C	*Put some water on the carrot.*
10	(450)	CI	*Put the carrot in the water.*
11	(552)	PC	*Pour the milk in the cereal.*
12	(557)	C	*Pour the cereal in the milk.*
13	(486)	C	*Pour the Coke in the lemonade.*
14	(488)	C	*Pour the lemonade in the Coke.*
15	(507)	C3	*Pour the juice in the egg.*
16	(510)	C	*Put the egg in the juice.*
17	(444)	C	*Put the rock in the water.*
18	(447)	PC	*Pour the water on the rock.*
19	(542)	C	*Put the raisins in the water.*
20	(546)	C2	*Pour some water on the raisins.*
21	(465)	PC	*Put the melon [peaches] in the tomatoes.*
22	(461)	C3	*Put the tomatoes in the melon.*
23	(451)	C	*Put the milk in the water.*
	(456)	C2	*Put the milk in the water*
24	(478)	C	*Pour the Perrier water in the milk.*
25	(525)	C	*Put the tomato in the oil.*
26	(528)	C	*Put some [the] oil in the tomato.*
27	(634)	I	*Put the shoe in the raisins.*
28	(543)	C2	*Put the raisins in the shoe.*
29	(502)	C	*Pour the juice in the Jello.*
30	(499)	C	*Open the Jello and pour it in the juice.*
31	(643)	C	*Rose/Nathaniel is gonna chase Kanzi/Alia.*
32	(636)	PC	*Kanzi/Alia is going to chase Rose/Mom.*
33	(635)	C	*Liz/Linda is going to tickle Kanzi/Alia.*
34	(655)	PC	*Kanzi/Alia is gonna tickle Liz/Nathaniel.*
35	(648)	C	*Kanzi/Alia is going to chase Liz/Nathaniel.*
36	(631)	C	*Liz/Nathaniel is going to chase Kanzi/Alia.*
37	(651)	C	*Kanzi/Alia is going to tickle Liz/Nathaniel with the bunny.*
38	(660)	PC	*Liz/Nathaniel is going to tickle Kanzi/Alia with the bunny.*
39	(580)	C	*Make the doggie bite the snake.*
40	(585)	C	*Make the snake bite the doggie.*
41	(330)	PC	*Hide the ball under the blanket.*
42	(600)	C	*Can you put the blanket on your ball?*

Table 4: Paired requests

3.2 Scoring

Although scoring Kanzi's full compliance or non-compliance is straight-forward, human interpretation is involved in between these two poles. Correct is different from Partially Correct. Partially Correct means that Kanzi attempted to comply with the whole request, but succeeded only in part. There are also degrees of correctness. Discussion of the grounds of decisions of this sort is found in the appendix (Savage-Rumbaugh et al., 1993: 111–210). For example, request (10) is scored as correct on the following grounds (Savage-Rumbaugh et al., 1993: 162):

> 450. (Cl) *Put the carrot in the water.* (Kanzi picks up a carrot, makes a sound like 'carrot,' takes a bite of the carrot, then puts it in the water.) [Cl is scored because Kanzi eats some of the carrot before putting it in the water.]

From our point of view, Kanzi is clearly treating the carrot as a Goal and as a Medium involved in the Minor Process of locating. He happens to eat a bit of it on the way to its destination, but he is not interpreting the request as simply a Goal-directed action, e.g. 'Eat the carrot'.

An example of human interpretation in scoring is request (23), when an unanticipated difficulty arises (Savage-Rumbaugh et al., 1993: 162):

> 451. (C) *Put the milk in the water.* (Kanzi picks up a closed can of SMA [milk], looks at the water, and shakes the milk, trying to figure out how to get the milk out of the can into the water.) E says, 'Put the milk, just put the whole can in the water.' (Kanzi looks around for something to open the milk with.) E says, 'Just put the can in, just drop the milk in the water.' [C is scored because Kanzi's behavior indicates that he has understood the sentence and is trying to figure out how to open the can so that he can pour the milk in the water. E's suggestions that he just put the whole can in the water are ignored, probably because, in his experience, the cans of SMA are opened and mixed with water, never just dropped in a bowl of water. Placing a can of milk in a bowl of water seems to make no sense to Kanzi.]

It seems that, for Kanzi, unlike the experimenter, 'the milk' indexes the liquid inside the closed can. The consequence is that although he can interpret the request, he cannot comply with it.

As a result of the difficulty, request (23) was repeated, but with a change (Savage-Rumbaugh et al., 1993: 164):

456. (C2) *Put the milk in the water*. (Kanzi is still poking the tomato
with his thumb from trial 455 while he listens to the sentence. After the
sentence, he picks up that tomato and puts it in the water.) E says, 'Put
the milk in the water.' (Kanzi pours the milk in the water.) [This is a representation of trial 451 to determine what Kanzi will do if the container
of milk is open, rather than closed, when the sentence is presented. Note
that Kanzi's continued interest in the item mentioned in the previous
trial is very similar to Alia's behavior on trial 451. The fact that Kanzi
now pours the milk directly into the water validates the interpretation of
the difficulty he encountered on trial 451.]

Both times, Kanzi can be seen to have interpreted the request as a Goal-
directed action modulating into a relational state, as in the previous examples
of pouring Coke in the lemonade and vice versa. Kanzi's interpretation of the
reverse request (24) 'Pour the Perrier water in the milk' is unproblematic.

4   Deacon's referential hierarchy: icon, index, and symbol

Distinguishing Kanzi's ability to interpret a command from his ability to
comply with it brings us to Deacon (1997). For Deacon, icons, indices,
and symbols are the result of three different interpretive processes made
possible by brain circuitry. Icons are recognitions. The key here is that an
icon is achieved when the interpretive process stops making distinctions.
In Deacon's (1997: 75–6) example, the hungry bird doesn't distinguish
between the bark of the tree and the camouflaged moth at rest on the bark.
That is, the bird has the potential to differentiate between edible and non-
edible objects, but is fooled in this particular case by the moth's not moving.
Indices involve a quite different interpretive process, the correlation of
icons, as in the familiar case of smoke as an index of fire. Symbols involve
yet another interpretive process, the discovery of relations between indices.
Instead of indexical Token-Object correlations, symbols involve Token-
Token relations.

Kanzi is clearly able to make iconic interpretations. In the case of 'Coke',
for example, Kanzi can make at least four different iconic interpretations.
He can recognize a can of Coke whether he sees it directly or in picture.
In so doing, Kanzi is ignoring the differences between the two sensory
images. Second, in recognizing abstract lexigrams, Kanzi ignores differences between the lexigram for Coke printed on a computer keyboard or
on large plastic sheet. Third, in recognizing the spoken word Coke, he
ignores differences of voice quality when the word is spoken by different

caretakers and researchers. Forth, there is of course the taste of Coke, which is distinguished from water or milk.

Equipped with these four icons, Kanzi is capable of making indexical interpretations: in this case, three closely related indices. The can of Coke icon is correlated with the taste icon; the lexigram for Coke is also correlated with the taste icon; and the spoken word icon is correlated with the taste icon. That is, the can, the spoken word, and the lexigram all predict the taste. Other correlations between these icons also hold. When the lexigram key is pressed, the computer speaks the word Coke: the lexigram predicts the spoken word. And either the spoken word or the lexigram can predict the can (weight, feel, temperature). In the data under consideration here, there are 42 such indexical signs.

Iconic interpretation is the foundation of indexical interpretation, and iconic and indexical interpretation form the foundation for symbolic interpretation. Symbolic interpretation, unlike the correlations of indexical interpretation, requires the ability to recognize combinatorial relations between indices. The big question is the extent to which Kanzi interprets the relations between indexical signs.

4.1 Icons and systems

What are 'relations' between indices, exactly? In order to answer this question, we have to step back and look at icons again. As previously mentioned, icons start where differentiation stops. The hungry bird stopped differentiating because the camouflaged moth didn't move. But this implies that the moth could have been differentiated from the bark of the tree, i.e. for the hungry bird there is a potential difference between edible objects and inedible objects. In other words, the hungry bird classifies or categorizes experience, and we can model this as a system, as in Figure 8.

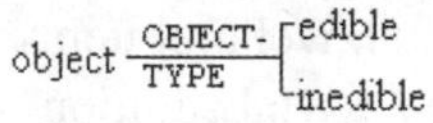

$$\text{object} \xrightarrow{\text{OBJECT-TYPE}} \begin{cases} \text{edible} \\ \text{inedible} \end{cases}$$

Figure 8: Iconic differentiation

This is, of course, a taxonomy. But it is also the template from which all systems derive because it models choice from a set of terms, or features, which have an entry condition, or a higher-order contextualization. That is, the system is the essence of the symbolic relation. But as Thibault (1997: 175), following Saussure, points out, it is difficult to derive a system from its terms. In human ontogeny, for example, children are motivated to acquire basic level terms, such as 'cat' and 'dog', in a folk taxonomy before they acquire superordinate terms, such as 'animal' (Halliday and Matthiessen, 1999: 83–9). But once entry conditions, or higher-order contextualizations are recognized, they make it possible to think symbolically as well as indexically.

4.2 Symbols and systems

Deacon's (1997: 90) account of how the chimpanzees Sherman and Austin moved from indexical interpretation to symbolic interpretation shows what happens when the pattern of iconic differentiation is used to model symbolic relations. In one experiment, Sherman and Austin (who had begun to discover symbolic relations), and Lana (who had experience only of indexical reference) learned to sort different kinds of food together in one bin and different tools together in another bin. In a subsequent experiment, Sherman and Austin succeeded, whereas Lana failed:

> This sorting task was followed by a second task in which the chimps were required to associate each of the previously distinguished food items with the same lexigram (glossed as 'food' by the experimenters) and each of the tool items with another lexigram ('tool'). Initially, this task simply required the chimps to extend their prior associations with bins to two additional stimuli, the two lexigrams. Although all three chimps learned his task in a similar way, taking many hundreds of trials to make the transference, Sherman and Austin later spontaneously recoded this information in a way that Lana did not. This was demonstrated when, as in the prior task, novel food and novel tool items were introduced. Sherman and Austin found this to be a trivial addition and easily guessed without additional learning which lexigram was appropriate.

Figure 9 models the symbolic relation between lexigrams.

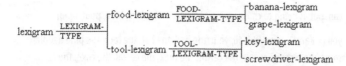

Figure 9: Sherman and Austin's symbolic differentiation

Because they have made explicit not only the terms of the system but the superordinate term, the higher order contextualization, Sherman and Austin are able to recode indexical interpretation as symbolic interpretation. Deacon (1997: 92) sums up Sherman and Austin's semiotic breakthrough this way:

> Sherman and Austin, as a result of their experience with a previous symbol system, recoded these new lexigram-object associations into two new symbolic categories that superseded the individual associations. It took them hundreds or thousands of trials to learn the first simple one-to-many associations. This was because they began with no systemic relationship in their still small lexigram repertoire for a general reference to 'food' or 'tool'. They had to learn them the hard way, so to speak, indexically. But as soon as they did learn these associations, they were primed to look for another higher-order logic, and once it was discovered, they were able to use this logic to generalize to new associations. Instead of hundreds or even thousands of trials, the availability of a symbolic coding allow them to bypass further trials altogether, an incredible increase in learning efficiency. The chimps essentially knew something that they had never explicitly learned. They had gained a kind of implicit knowledge as a spontaneous byproduct of symbolic recoding.

4.3 Kanzi's indexical signs

From the 21 pairs of requests, we identify 42 types of spoken indexical signs, with tokens ranging from one occurrence (e.g. 'under', 'it') to 22 occurrences (e.g. 'put', 'in'). The number of types is within the range of vervet monkeys calls (Cheney & Seyfarth, 1990), which are themselves indexical signs. Table 5 displays the inventory of indexical signs as types of word groups.

Verbal group	can, put, is going to chase, going to tickle, make, bite, hide, pour, open
Nominal group	you (=Kanzi), some/the oil, your/the ball, the hat, the rock, the pine needles, some/the [Perrier] water, the carrot, the milk, the cereal, the coke, the lemonade, the juice, the egg, the raisins, the melon, the tomato(es), the shoe, the jello, it(=jello), Rose, Kanzi, Liz, the bunny, the doggie, the snake, the blanket
Preposition group	on, in, with, under
Conjunction	And

Table 5: Indexical sign types of paired requests

More than word order is involved in the question of how these indexical signs are related. And of course the symbol system that relates them doesn't cancel out their indexicality. Indeed, it is the clause rather than word that is at issue, since Kanzi's interpretations necessarily involve the representation of participants, processes, and circumstances.

5   Kanzi's grammar

Wu (2000: 37–8) provides a clear statement of basic systemic functional theory in general, and of the clause in particular:

> Language is interpreted in systemic functional grammar as a resource.
> The resource is organized along two axes: the paradigmatic and the
> syntagmatic. Paradigmatic and syntagmatic organization are modeled
> by different representational resources. The paradigmatic organization
> is represented by system networks (networks of interrelated systems of
> options), and the syntagmatic organization is represented by function
> structures (configurations of constituent functions). The paradigmatic
> organization is primary and the syntagmatic organization is secondary;
> syntagmatic relations are derivable from paradigmatic selections by
> means of realization statements.

Hence, we will separate the paradigmatic organization from the syntagmatic organization, and treat the ordered indexical signs as being derivable from paradigmatic selections.

5.1 Paradigmatic representation

There are two ways of going about this. One is to make as complete a representation of meaning potential as possible (as for example in Matthiessen, 1995), and to see what parts of the over-all system Kanzi appears to be processing. The other is to make a minimal system which would account for Kanzi's ability to process indices as symbols. For now, we're taking the second route, with the system we're about to propose. Part of the minimalism is that we're only looking at the ideational metafunction, although interpersonally Kanzi clearly distinguishes between the exchange of goods and services and information. He can make and comply with commands, and he can respond to both information seeking and polar questions (Taglialatela et al., 2004).

Wu (2000: 39), following Matthiessen (1995: 235), proposes a transitivity network for the clause which has two simultaneous systems: PROCESS-TYPE and AGENCY. We are proposing a similar network, which replaces the AGENCY system with a PARTICIPATION-TYPE system (since the data do not require differentiation between middle and effective voice), and adds a CIRCUMSTANTIATION-TYPE system. Figure 10 displays this network.

Figure 10: Kanzi's symbolic relations between indices

There are both similarities and differences between Figure 9 (Sherman and Austin's symbolic relations between lexigrams) and Figure 10 (Kanzi's symbolic relations between indices). The main similarity is that both systems contain features and entry conditions. In Figure 9, 'lexigram' is the entry condition for a system forcing a choice between the features 'food-lexigram' or 'tool-lexigram', with each of these features the entry conditions for further systems. In Figure 10, 'location' is the entry condition forcing a choice between 'positive-position' and 'relative-position'. In the network diagrams square brackets indicate such 'or' choices.

The main difference between the figures is that in Figure 10, the initial entry condition, 'clause', forces choice in three simultaneous systems:

'PARTICIPATION-TYPE', 'PROCESS-TYPE', and 'CIRCUMSTANTIATION-TYPE'.
In the network diagrams curly brackets indicate such 'and' relations.

5.1.1 *Traversing the network*

Wu (2000: 41–2) goes on to say:

> The system network, as a representation of paradigmatic relations,
> does not specify grammatical structures; it only specifies grammatical
> paradigmatic specifications by means of system traversal…. In other
> words, the result of each traversal through the network is simply the
> complete set of grammatical features that distinguishes one linguistic
> unit from all others described by the system network.

If we traverse the network in Figure 10, we get 'a set of grammatical features
that distinguishes one linguistic unit from all others described by the system
network'. The network allows eight distinct traversals, as follows:

(1) **clause: two-participant-process & doing & minor-process:
    location: positive-position**

(2) **clause: two-participant-process & doing & minor-process:
    location: relative-position**

(3) **clause: two-participant-process & doing & minor-process:
    means**

(4) **clause: two-participant-process & doing &
    no-circumstantiation**

(5) clause: three-participant-process & doing & minor-process:
    location: positive-position

(6) clause: three-participant-process & doing & minor-process:
    location: relative-position

(7) clause: three-participant-process & doing & minor-process:
    means

(8) **clause: three-participant-process & doing &
    no-circumstantiation**

These traversals represent the meaning potential of this system. The unique
combination of features of each traversal is what makes the system symbolic. As it happens, only five of the possible combinations, those shown
in bold, are actually represented in the data.

5.2 Syntagmatic representation

Wu (2000: 41–2) continues:

> To produce grammatical structures we need to appeal to realization statements, which are specified in the context of systemic features… It is in the systemic contexts that grammatical structures are specified. More specifically, all realization statements are specified in the context of systemic features. For example, the realization statement '+Subject' is specified in the context of the feature 'indicative' while '+Actor' in the context of the feature 'material'.

Figure 11 shows the network of Figure 10 with realizations added.

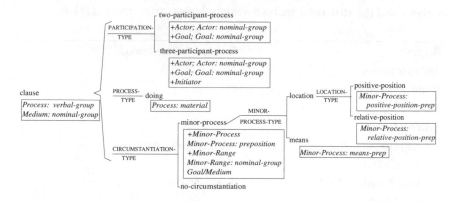

Figure 11: System network with realizations

Thus, for example, in the systemic environment 'two-participant-process' (in the PARTICIPATION-TYPE system) the presence of an Actor is required (+Actor), and the Actor is required to be filled by a nominal group (Actor: nominal-group). Similarly, a Goal is required, and its filler specified as being also a nominal-group. In the 'three-participant-process' environment, an Initiator function is also specified. In the environment of minor-process (in the CIRCUMSTANTIATION-TYPE system), a Minor-Process is additionally required, preselected to be a preposition. The exact type of the preposition is specified further in the dependent systems, which specify the Minor-Process to be either a positive-position prep (*in, on*), a relative-position-prep (*under*) and means-prep (*with*). Note also that in the environment of a minor process, the conflation of the functions of Goal and Medium is specified.

5.3 Systemic and structural analysis

We are now in a position to make both a systemic and a structural analysis of the 42 requests. These fall into five traversals of Kanzi's network. We will discuss some examples for each traversal.

5.3.1 Traversal (1)

> Clause: two-participant-process & doing & minor-process: location: positive-position (e.g. 'Pour the Coke in the lemonade' vs. 'Pour the lemonade in the Coke').

Traversal (1) corresponds to request-type 1A: *Put object X in/on transportable object Y* (Savage-Rumbaugh et al., 1993). Tables 6 displays both the systemic analysis and the structural analysis which derives from requests (13).

Request:	*Pour the Coke in the lemonade*
Table 8 sequence:	13
Running sequence:	486
Score:	C
Systemic Analysis:	
PARTICIPATION TYPE	two-participant-process
PROCESS TYPE	doing
CIRCUMSTANTIATION TYPE	positive-position
Structural Analysis:	
Process:material	pour
Actor	(Kanzi)
Goal/Medium	the Coke
Minor Process: location: positive-position	in
MinorRange	the lemonade

Table 6: Analysis of *Pour the Coke in the lemonade*

The analysis represents that part of the human symbol system that Kanzi is processing. By doing what was requested, Kanzi shows that he has interpreted the string of indices symbolically, by mapping them onto the syntagmatic functions, which in turn are derived from the paradigmatic system. In (13), the indexical sign *pour* is mapped onto the function-structure material process, which requires an Actor, i.e. Kanzi. The indexical sign *the Coke* is mapped onto the function-structure Goal; that is, Kanzi's interpretation is that he performs the Actor function, and that his material process of pouring extends to a Goal. But there are other indexical signs. The indexical sign *in* is mapped onto the function-structure Location. This means that the previous Goal becomes the Medium of the Minor-Process Location. Finally, the indexical sign *the lemonade* is mapped onto the function-structure Minor-Range, with the result that the Goal of his action will come to a state of rest in a particular place. In other words, in addition to making iconic and indexical interpretations, Kanzi is making a symbolic interpretation of the request.

Request:	*Pour the lemonade in the Coke*
Table 8 sequence:	14
Running sequence:	488
Score:	C
Systemic Analysis:	
PARTICIPATION TYPE	two-participant-process
PROCESS TYPE	doing
CIRCUMSTANTIATION TYPE	positive-position
Structural Analysis:	
Process:material	pour
Actor	(Kanzi)
Goal/Medium	the lemonade
Minor Process: location: positive-position	in
MinorRange	the Coke

Table 7: Analysis of *Pour the lemonade in the Coke*

Table 7 shows the analysis of request (14), which reverses the two liquids. In this request, precisely the same symbolic relations of system-choice and structural realization hold, but different indexical signs are mapped on to them. The indices remain indices, but their relations are systematically different.

Traversal (1) includes most of the requests: 1–30 and 42. Traversal (2) is not discussed here because of its similarity to Traversal (1) items. Traversal (2) includes only request 41.

5.3.2 *Traversal (3)*

> Clause: two-participant-process & doing & minor-process: means
> (e.g. 'Kanzi is going to tickle Liz with the bunny' vs. 'Liz is
> gonna tickle Kanzi with the bunny').

Traversal (3) corresponds to request-type 4: *announce information* (Savage-Rumbaugh et al., 1993). Request (37), 'Kanzi is going to tickle Liz with the bunny' involves a different kind of CIRCUMSTANTIATION than the previous examined, but is also interpreted without problem by Kanzi. The systemic and structural analysis representing that part of the human symbol system Kanzi is processing is shown in Table 8.

Request:	*Kanzi is going to tickle Liz with the bunny*
Table 8 sequence:	37
Running sequence:	651
Score:	C
Systemic Analysis:	
PARTICIPATION TYPE	two-participant-process
PROCESS TYPE	doing
CIRCUMSTANTIATION TYPE	means
Structural Analysis:	
Process:material	tickle
Actor	Kanzi
Goal	Liz
Minor Process: means	with
MinorRange	the bunny

Table 8: Analysis of *Kanzi is going to tickle Liz with the bunny*

(Savage-Rumbaugh et al., 1993: 208) give this account of Kanzi's compliance:

> 651. (C) Kanzi is going to tickle Liz with the bunny. (Kanzi picks up the
> bunny puppet, puts it on his hand, walks over to Liz, and begins tickling
> her leg. He also tickles Linda.) E says, 'Just Liz.' (Kanzi returns to
> tickling Liz.)

Here the similarity between the preposition and the corresponding verb is very clear (cf. Halliday, 1994: 158: 'he cleaned the floor with a mop ~ using a mop').

A new difficulty, different from that of request (23), 'Put the milk in the water' discussed in Section 3.2, arises with request (38) (Savage-Rumbaugh et al., 1993: 210), which is scored not as Correct, but as Partially Correct:

> 660. (PC) *Liz is gonna tickle Kanzi with the bunny.* (Kanzi stands up and lifts the toy gorilla up briefly. It is laying on top of his keyboard.) E says, 'Give Liz the bunny so she can tickle you.' (Kanzi takes the toy gorilla to Liz and drops it on the floor in front of her with a play face. Liz picks it up and begins to tickle Kanzi.)

As mentioned in Section 2.2, Partially Correct means that Kanzi attempted to comply with the request but succeeded only in part; in this particular case, he has picked up the toy gorilla rather than the bunny. More interesting from our point of view is the fact that request (38) is 'announcing information', but it is semantically a request directed not at Kanzi but at Liz, who can't hear it because of the music in her headphones. Kanzi is not the Actor in the process of tickling, so he doesn't act. When the experimenter says 'Give Liz the bunny so she can tickle you', this is a request directed at Kanzi, not at Liz, and this time he does act. Kanzi complies with the request with a Goal-directed action. Kanzi's 'playful face' suggests that he remembers the previous 'Liz is gonna tickle Kanzi with the bunny'. In other words, by not initiating action, until E's direct command, Kanzi is behaving as appropriately in request 38, and he did in request 37, when he was the Actor in a process of tickling with an instrument.

Traversal (3) includes only requests 37 and 38.

5.3.3 Traversal (4)

> Clause: two-participant-process & doing & no-circumstantiation (e.g. 'Kanzi is going to chase Rose' vs. 'Rose is going to chase Kanzi').

This too corresponds with type 4 requests: *announce information*. Table 9 displays both systemic analysis and the structural analysis which derives from it for request (32).

Request:	*Kanzi is going to chase Rose*
Table 8 sequence:	32
Running sequence:	636
Score:	PC
Systemic Analysis:	
PARTICIPATION TYPE	two-participant-process
PROCESS TYPE	doing
CIRCUMSTANTIATION TYPE	none
Structural Analysis:	
Process: material	is going to chase
Actor	Kanzi
Goal	Rose

Table 9: Analysis of *Kanzi is going to chase Rose*

Kanzi's interpretation of request (32) shows the difficulty of complying with a chasee who doesn't run away, presumably because of the headphones (Savage-Rumbaugh et al., 1993: 205):

> 636. (PC) *Kanzi is going to chase Rose.* (Kanzi looks at Rose and scoots over toward her, as though waiting for her to run away. Rose does nothing. Kanzi touches Rose. Rose gets up, and Kanzi then backs away, stops, looks at Rose, and waits for her to run. Rose doesn't, so Kanzi approaches instead.) *Error correction. E* tells Rose what is supposed to happen. Kanzi then gestures toward Rose, and she chases him.

Once again, it takes two to tango. When Rose doesn't exhibit chase behaviour, Kanzi's compliance is thwarted because Rose cannot hear the request. Although Rose is told what is supposed to happen, she interprets Kanzi's gesture as an invitation to chase him.

On the other hand, Kanzi's interpretation of request (31) overcomes the difficulty of the caretaker not being able to hear what she is supposed to do in a joint action. Table 10 displays the systemic and structural analysis.

Request:	Rose is going to chase Kanzi
Table 8 sequence:	31
Running sequence:	643
Score:	C
Systemic Analysis:	
PARTICIPATION TYPE	two-participant-process
PROCESS TYPE	doing
CIRCUMSTANTIATION TYPE	none
Structural Analysis:	
Process: material	is going to chase
Actor	Rose
Goal	Kanzi

Table 10: Analysis of *Rose is going to chase Kanzi*

(Savage-Rumbaugh et al., 1993: 205) describe what happens:

> 643. (C) *Rose is gonna chase Kanzi.* (Kanzi looks at Rose.) E says,
> 'Rose is going to chase you.' (Kanzi looks at Rose, puts his bowl down,
> signs *chase*, points to Rose, then runs away. Rose chases Kanzi.)

Kanzi interprets the experimenter's 'announcement of information' correctly, but he has to bring it into being, to set this Goal-directed action into motion. Kanzi achieves this by doing three different yet integrated multimodal things: (1) he nominates Rose as Actor by gesture; (2) he indicates the process by signing chase (whether by clapping his hands together – a conventional signal among the bonobos at the Language Research Center, or by pointing to the CHASE lexigram); and (3) he makes himself the Goal of Rose's action by running away.

Traversal (4) includes requests 31, 32, 33, 34, 35, and 36.

5.3.4 Traversal (8)

> Clause: three-participant-process & doing & no-circumstantiation
> (e.g. 'Make the doggie bite the snake' vs. 'Make the snake bite
> the doggie').

This traversal corresponds to type 6 request: *Make pretend animate* A *do action* A *on recipient* Y. Table 11 displays the systemic and structural analysis of request (39).

Request:	Make the doggie bite the snake
Table 8 sequence:	39
Running sequence:	580
Score:	C
Systemic Analysis:	
PARTICIPATION TYPE	three-participant-process
PROCESS TYPE	doing
CIRCUMSTANTIATION TYPE	none
Structural Analysis:	
Process: material	make… bite
Initiator	(Kanzi)
Actor	the doggie
Goal	the snake

Table 11: Analysis of *Make the doggie bite the snake*

Savage-Rumbaugh et al. (1993: 192) discuss Kanzi's compliance as follows:

> 580. (C) *Make the doggie bite the snake.* (Kanzi picks up the dog and puts it on the snake, then moves it back, picks up the snake, and looks at its mouth.) E says, 'Make the doggie bite the snake.' (Kanzi puts the snake's mouth up to the doggie's mouth.) E says, 'Yeah, that's right. Un huh. Thank you.' (Kanzi opens the dog's mouth and sticks the snake's head in the dog's mouth.) E says, 'Yeah, push his mouth down. Yeah, that's very good, Kanzi.' (Kanzi pulls the snake back and puts it down.) [C is scored because Kanzi does not hesitate at any point and his actions appear to be directed smoothly toward carrying out the request.]

Here Kanzi is clearly interpreting the request as requiring that he cause the inanimate objects to behave as if they were animate, i.e. to position the doggie as carrying out Goal-directed action.

Similarly, the systemic and structural analysis of the reverse request (40) is displayed in Table 12.

Request:	Make the snake bite the doggie
Table 8 sequence:	40
Running sequence:	585
Score:	C

Systemic Analysis:

PARTICIPATION TYPE	three-participant-process
PROCESS TYPE	doing
CIRCUMSTANTIATION TYPE	none

Structural Analysis:

Process: material	make... bite
Initiator	(Kanzi)
Actor	the snake
Goal	the doggie

Table 12: Analysis of *Make the snake bite the doggie*

Savage-Rumbaugh et al. (1993: 193) provide the following account of Kanzi's compliance:

> 585. (C) *Make the snake bite the doggie.* (Kanzi picks up the snake and then the dog. Kanzi pushes the snake's mouth down onto the dog's mouth.) E says, 'Uh huh, that's real good.' (Kanzi holds the snake's mouth on the doggie's mouth.)

Once again, Kanzi is clearly interpreting the request as requiring that he cause the inanimate same objects to behave as if they were animate, but in this case it is the snake that is made to engage in Goal-directed action.

Traversal (8) includes only requests 39 and 40.

6 Conclusion

We have represented the experimenter's 42 requests from a systemic perspective. We have constructed a system network, and associated structural realizations with it (the closest thing to 'syntax' in systemic theory). We have treated the requests as consisting of a series of indexical signs, in the spirit of Deacon's referential hierarchy, and we have shown that the proposed system shows precisely how these indexical signs are symbolically related. We then looked at the discussion from Savage-Rumbaugh et al., (1993) of Kanzi's compliance with the requests as a function of his interpretation of what he hears. For Kanzi, much more than word order is involved in his interpretations. Since Kanzi's interpretations can plausibly be said to be based on symbolic relations between indices, there is evidence that a bonobo brain can process human symbolic language at least to the extent described here, and thus manifests some degree of human-like consciousness.

Acknowledgement

This article is a revised version of a paper given at the Second International Workshop of the Systemic Functional Research Community on Interpersonal and Ideational Grammar (University of Leuven, 21–24 November 2001). The Research Community is funded by the Fund for Scientific Research – Flanders (Belgium) (grant no. WO.018.00N).

Note

1   http://www.wagsoft.com/Systemics/

References

Calvin, W. H. and Bickerton, D. (2000) *Lingua Ex Machina*: *reconciling Darwin and Chomsky with the human brain*. Cambridge, MA and London: MIT Press.

Cheney, D. L. and Seyfarth, R. M. (1990) *How Monkeys See the World*: *inside the mind of another species* Chicago: University of Chicago Press.

Damasio, A. R. (1999) *The Feeling of What Happens*: *body and emotion in the making of consciousness*. New York: Harcourt Brace.

Davidse, K. (1992) Transitivity/ergativity: the Janus-headed grammar of actions and events. In M. Davies and L. Ravelli (eds) *Advances in Systemic Linguistics*. London and New York: Pinter.

Deacon, T. W. (1997) *The Symbolic Species*: *the co-evolution of language and the brain*. London: W.W. Norton.

Edelman, G. and Tononi, G. (2000) *A Universe of Consciousness*: *how matter becomes imagination*. New York: Basic Books.

Eggins, S. (1994) *An Introduction to Systemic Functional Linguistics*. London: Frances Pinter.

Eggins, S. and Slade, D. (1997) *Analysing Casual Conversation*. London: Cassell.

Fouts, R. and Waters, G. (2001) Chimpanzee sign language and Darwinian continuity: evidence for a neurological continuity for language. *Neurological Research* 23: 787–94.

Halliday, M. A. K. (1994) *An Introduction to Functional Grammar*. London: Edward Arnold.

Halliday, M. A. K. (2001) Meanings, wordings and contexts: modelling the 'language brain'. Paper delivered at Brain Sciences Institute, RIKEN.

Halliday, M. A. K. and Matthiessen, C. M. I. M. (1999) *Construing Experience Through Meaning: a language-based approach to cognition.* London: Cassell.

Martin, J. R., Matthiessen, C. M. I. M. and Painter, C. (1997) *Working with Functional Grammar.* London: Edward Arnold.

Matthiessen, C. M. I. M. (1995) *Lexicogrammatical Cartography: English systems.* Tokyo: International Language Sciences Publishers.

Savage-Rumbaugh, S., Murphy, J., Sevcik, R., Brakke, K., Williams, S. and Rumbaugh, D. (1993) *Language Comprehension in Ape and Child.* Monographs of the Society for Research in Child Development. Chicago: University of Chicago Press.

Taglialatela, J., Savage-Rumbaugh, S., Rumbaugh, D. M., Benson, J. and Greaves, W. (2004) Language, apes, and meaning-making. In G. Williams and A. Lukin (eds) *The Development of Language: functional perspectives on species and individuals.* London: Continuum.

Thibault, P. (1997) *Re-reading Saussure: the dynamics of signs in social life.* London and New York: Routledge.

Thompson, G. (1996) *Introducing Functional Grammar.* London: Edward Arnold.

Wu, C. (2000) *Modelling Linguistic Resources: a systemic functional approach.* Macquarie University: Linguistic PhD thesis.

3   The evolutionary dimension: the thin end of the wedge – grammar and discourse in the evolution of language

James D. Benson, William S. Greaves, Sue Savage-Rumbaugh, Jared P. Taglialatela and Paul J. Thibault

Abstract

The human language system is both multifunctional and multistratal (Halliday, 1994). The lexicogrammatical stratum both differentiates and integrates three metafunctional systems: ideational, interpersonal and textual. Expression systems are 'below', and semantic systems 'above' lexicogrammatical systems. In contrast, indexical primate call systems (Deacon, 1997) are without a metafunctionally differentiated lexicogrammar slotted in between semantics and expression. In trying to explain how this gap could be bridged, we examine three snapshots taken from comprehensive studies with large amounts of data. First, we look at evidence for proto-metafunctional differentiation in monkeys (Zuberbühler, 2005). Second, we look at a evidence in a human child for the transition from indexical call systems to ideationally and interpersonally differentiated symbolic lexicogrammar (Halliday, 1975; 1995). Third, we look at evidence in a bonobo-human interaction (Savage-Rumbaugh et al., 1993; Savage-Rumbaugh & Lewin, 1994; Savage-Rumbaugh, Shanker & Taylor, 1998) for the interpretation of ideationally and interpersonally differentiated human lexicogrammar, and for the recruitment of bonobo vocalizations as a means for ape participation in the human discourse-semantics system on the basis of lexicogrammatical interpretation. That an ape brain is capable of this degree of symbolic exchange of information with a human interactant suggests that the development of social discourse played an important part in bridging the gap between indexical calls and symbolic human language.

1. Introduction: grounding assumptions

1.1 Metafunctions and strata

The intricacies of human language reach far beyond syntactic complexity. In order to capture this complexity, we make a number of assumptions about language and the social context of which it is a part; these allow us to compare three snapshots selected from comprehensive studies with large amounts of data, from an evolutionary perspective: monkey-monkey interactions, human child-caregiver interactions, and ape-human interactions.

Language complexity can be represented by two axes: metafunction and stratification (Fawcett et al., 1993; Halliday, 1994; Halliday & Matthiessen, 1999; Martin, 1992; Matthiessen, 1995; Wu, 2000).

The metafunctional axis is divided into three different modes of meaning: interpersonal, ideational, and textual. The interpersonal metafunction is concerned with *enacting* interpersonal relations, distinguishing, for example, giving or demanding information from giving or demanding goods and services. The ideational metafunction is concerned with *representing* processes, participants, and circumstances, distinguishing, for example, goal-directed actions from relational processes, or circumstances of location from circumstances of manner. The textual metafunction is concerned with *organizing* combined interpersonal and ideational meanings into messages that cohere with context, for example, distinguishing given from new information in an utterance.

The layers of the stratification axis are: context, meaning, wording, and sounding. The top layer, context, involves three types of behaviour: interacting, doing, and communicating. These are correlated with the three different modes of meaning. Interactions between people are correlated with interpersonal meanings; doings involving material and non-material things of this world are correlated with ideational meanings; and communications through the creation of messages is correlated with textual meanings. The level below context is semantics (meaning), followed by lexicogrammar (wording), and phonology (sounding), or some other system of expression.

Figure 1 summarizes the two axes, and includes examples of interpersonal and textual systems operating at different levels.

	Context	Interacting	Doing	Communicating
Strata		Metafunctions: different modes of meaning		
		Interpersonal: enacting interpersonal relations	Ideational: representing experience	Textual: organizing message
	Semantics	e.g. speech functions: command, statement, question		e.g. salience vs. non-salience
	Lexicogrammar	e.g. mood system: imperative, declarative, interrogative		e.g. information system: given, new
	Expression e.g. phonology	e.g. falling / rising intonation		e.g. placement of an intonational fall or rise

Figure 1: Summary of the axes of metafunction and strata

1.2 Discussion of Figure 1

Context is divided into three types of behaviour: interacting, doing, and communicating, which are correlated with the interpersonal, ideational, and textual modes of meaning.

At the level of language, lexicogrammar is an abstract set of metafunction-ally different systems that mediate semantics and expression. The first example represents choices within the interpersonal metafunction. Choices from the speech function system (semantics) are expressed twice: first by choices from the mood system (lexicogrammar), and second by choices from the system of rises and falls (phonology). The second example represents choices within the textual metafunction. Choices from the salience system (semantics) are also expressed twice: first by choices from the given-new information system (lexicogrammar), and second by choices from the placement system of intonation (phonology).

2   Monkey calls: indexical signs and emergent semiotic complexity

2.1 Monkey calls as indexical signs

Unlike human language with its open potential for symbolic meaning, vervet monkey alarm calls (Cheney & Seyfarth, 1990) form a closed set of meaning-expression pairs. Such calls can be parsimoniously represented in Deacon's (1997) terms as indexical signs, i.e. correlations of icons, shown in Figure 2.

Indexical sign

Icon x   Icon y

Figure 2: Indexical signs as the correlation of icons

For example, there is a high probability that a vervet will make an eagle call (icon x) when an eagle (icon y) is sighted, and there is an equally high probability that other vervets will make echoic eagle calls, and will take anti-predator action appropriate to eagles, rather than to leopards. This suggests that they recognize the original eagle indexical sign. The number of indexical meanings is limited by the number of sounds the monkey can make. Deacon (1997) argues that such indexical signs are radically different from symbols, which result from combinatorial relations among indices.

2.2 Emergent semiotic complexity of monkey calls

The calls of Diana and Campbell's monkeys discussed by Zuberbühler (2005) are more semiotically complex than Figure 2 suggests. Both Diana monkeys and Campbell's monkeys produce eagle alarm vocalizations, although the two calls are acoustically quite different (Zuberbühler, 2002). Diana monkeys respond to Campbell's monkey alarm calls by taking the appropriate anti-predator action, and also by producing, as an echoic response their own Diana eagle alarm calls. This suggests that they have interpreted the Campbell's eagle alarm indexical sign, although they cannot reproduce it.

The Campbell's monkeys have a more complex system of calls, which the Diana monkeys also seem to be able to interpret. In addition to eagle and leopard alarm calls, Campbell's monkeys sometimes produce 'boom' calls. The 'boom' calls can occur by themselves, or precede alarm calls by 30 seconds. In the former case, the 'boom' calls seem to indicate a minor disturbance or a distant predator, but in the latter case, since they have already been alerted by the 'boom', echoic responses and correlated anti-predator behaviour by the Diana monkeys is greatly reduced.

The semiotic complexity of this behaviour is revealed by considering separately the different kinds of meanings that the Campbell's monkeys are making with these calls. First, there are meanings having to do with doing: the type of predator and type of anti-predator action. Second, there are meanings having to do with interacting: to whom are their calls directed? Third, there are meanings having to do with communicating: the mechanics of sending the message. That is, the meaning of the 'boom' call is distinct from both interactive relations between monkeys, and differentiating among predators; rather, the 'boom' call indicates whether or not the predator information is newsworthy, *i.e.* it structures a communicative message. In other words, the monkeys' indexical signs are beginning to show the differentiation of modes of meaning.

We can contrast the representation of the indexical sign in Figure 2 with a representation of this emerging semiotic complexity in Figure 3. Figure 3 is arranged in rows and columns. The rows in Figure 3 label three different strata: Context, or what the monkeys are doing behaviourally, Semantics, or what the monkeys are meaning, and Expression, or how the monkeys are sounding. The columns in Figure 3 label three different metafunctions: Proto-ideational, Proto-interpersonal, and Proto-textual.

2.2.1 Discussion of Figure 3: emergent metafunctional differentiation of Campbell's monkey calls

The context row in Figure 3 shows that the monkeys are doing three different things. First, they sort out and deal with what constitutes an external threat. They distinguish near predators from distant predators, they distinguish major threats from minor ones, and they are able to act appropriately on the basis of these perceptions. Second, they socially take turns in communicating with each other. Third, they can communicate the message in a way that makes the salience of a threat clear.

The semantics row of Figure 3 shows that the monkeys are making three different kinds of meaning which correlate with the three different kinds of things they are doing. The first column of the semantic row shows ideational meaning, the second shows interpersonal meaning, and the third shows textual meaning. Meanings are represented as systems, and the associated boxes represent the expression paired with that meaning.

Ideationally, their meaning system distinguishes between major dangers (eagle alarms, leopard alarms) and minor perturbations (alerts) of their environment. These meanings are paired with eagle, leopard, and boom expressions.

Interpersonally, by choosing among their relatively small inventory of calls they are able to structure a simple form of discourse. They can make initiating moves vs. sustaining moves, and they can respond in two ways, either by echoing an initiating call or by remaining silent. But, because they are constrained by this system, they cannot complicate the discourse by, for example, contradicting the initiating move of an eagle call with a leopard call.

C O N T E X T	Doing: Different predators Specific anti-predator routines Minor disturbance in the environment	Interacting: Vocalizations directed towards predators and/or conspecifics Turn-taking	Communicating: Response to salient aspects of the environment on the basis of vocalizations
S	Proto-ideational	Proto-interpersonal	Proto-textual

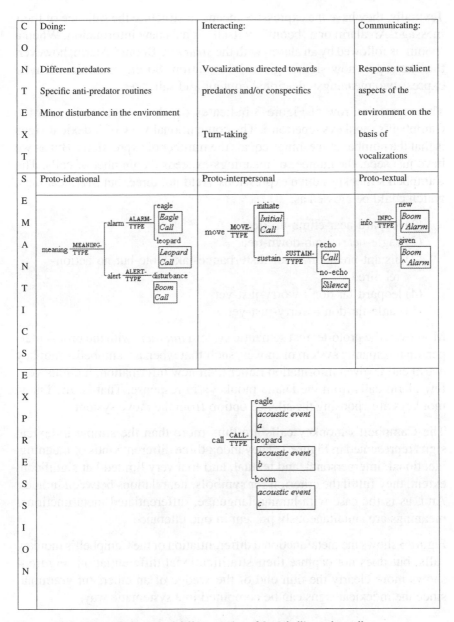

Figure 3: Emergent metafunctional differentiation of Campbell's monkey calls

Textually, they have the expressive means to structure the salience of their message. An alarm or a 'boom' by itself signals new information. When a 'boom' is followed by an alarm, with the *structure* Boom ^ Alarm, however, the alarm part now signals given information. So the one call will now express two meanings: ideational content and salience.

The expression row of Figure 3 indicates the three vocalizations in the Campbell's monkeys repertoire. The conventional view of indexical signs is that the number of meanings equals the number of expressions. But as we have just seen, the number of meanings exceeds the number of calls. The Campbell's monkeys three expressions yield not three, but *five* meanings, which could be glossed as:

(1) leopard-near-climb-up-tree
(2) eagle-near-climb-down-tree
(3) distant-predator/minor-disturbance-take-note-but-no-action-required
(4) leopard-far-don't-worry-just-yet
(5) eagle-far-don't-worry-just-yet.

Moreover, the proto-textual semantic system *interacts* with the proto-interpersonal semantic system of moves, such that when a Campbell's monkey alarm call is given information rather than new information, it elicits very few alarm calls from the Diana monkeys in response. That is, the Diana monkeys are choosing the silence option from the move system.

The Campbell's monkey calls are thus more than the simple indexical signs represented in Figure 2. They index three different kinds of meaning, ideational, interpersonal, and textual, and to a very limited but significant extent, they fulfill the criterion for symbols, i.e. relations between indices. And, as is the case with human language, differentiated metafunctional meanings are simultaneously present in one utterance.

Figure 3 shows the metafunctional differentiation of the Campbell's monkey calls, but does not capture their stratificational differentiation. Figure 4 shows more clearly the thin end of the wedge of an emergent grammar, since the indexical signs can be combined in a systematic way.

2.2.2 Discussion of Figure 4: grammar and emergent stratificational differentiation in Campbell's monkey calls

The semantics row of Figure 4 corresponds to the semantics row of Figure 3. Our semantic row does not include an interpersonal column. Zuberbühler (2005) argues that the interaction of 'boom' and alarm calls is interpersonal – a kind of hedging, but the authors see this as more directly related to the distribution of given-new information.

Semantics	Ideational: two types of specific danger (eagle or leopard) and one general danger (disturbance)	Textual: two kinds of information (salient and non-salient)
Grammar slotted in between semantics and expression curly brackets indicate simultaneous 'and' choices; square brackets indicate alternative 'or' choices	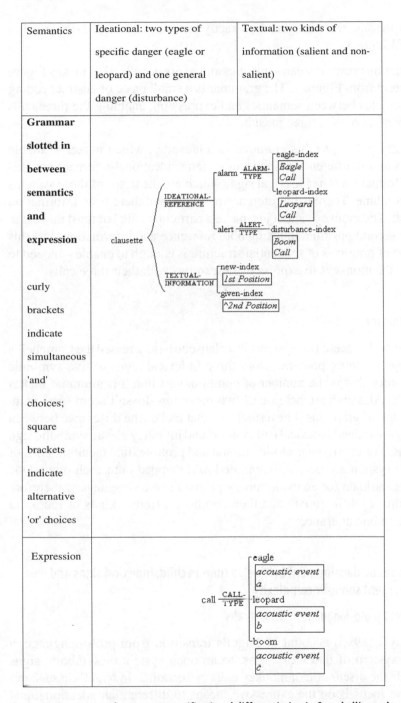	
Expression		

Figure 4: Grammar and emergent stratificational differentiation in Campbell's monkey calls

The expression row of Figure 4 is exactly the same as that of the expression row of Figure 3.

The grammar row between semantics and expression is what makes Figure 4 different from Figure 3. The grammar is a small piece of abstract coding that mediates between semantics and expression, enabling the three calls to convey more than three meanings.

The entry point into this grammar is 'clausette', which makes available choices in two different simultaneous systems: ideational-reference and textual-information. The indexical signs which are the terms of these systems must combine. There is no reference by itself, and there is no information by itself. The monkey cannot just make a particular call, but must make it in first or second position, indexing both reference and information. Just this little bit of grammar of sequential structure is enough to enable – indeed to force – the monkeys to express five meanings with their three calls.

2.3 Summary

Choices within these two systems in a clausette has increased the Campbell's monkeys meaning potential from three indexical signs to five symbolic signs, since that is the number of combinations that this grammar makes possible. Although an increase of two meanings doesn't seem very great, the wedge of grammar is nevertheless what makes the difference between highly constrained indexical call systems and infinitely elastic symbolic sign systems. The emergent proto-ideational and proto-textual metafunctions of the call system are *both* differentiated *and* mapped onto each other. One strong candidate for an evolutionary prerequisite to language is therefore the ability to differentiate and then combine different kinds of indexical signing in one utterance.

3   Language development in Nigel, a human child: indexical signs and emergent semiotic complexity

3.1 Nigel's proto-language: indexical calls

Halliday's (1995) account of Nigel's transition from proto-language, a closed system of indexical signs, to an open system of symbolic signs parallels the discussion of monkey calls in Section 2. In Nigel's case, however, the focus is on the expressive means to differentiate ideational and interpersonal meaning rather than ideational and textual meaning.

Among Nigel's proto-language signs were a set of calls: 'mummy', 'daddy', and 'anna', each of which indexed the appropriate caregiver. These meaning-expression pairs are less complex than those of the Campbell's monkey calls discussed in Section 2. The system of calls shown in Figure 5 is specifically concerned with the to-be-differentiated proto-ideational and proto-interpersonal metafunctions, rather than the differentiated proto-textual and proto-ideational systems discussed in Figure 4.

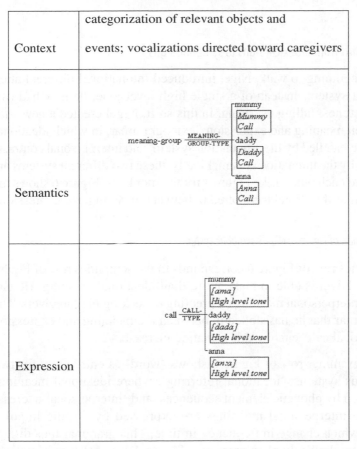

Figure 5: Nigel's indexical calls

3.1.1   Discussion of Figure 5: Nigel's indexical calls

The context row in Figure 5 shows what Nigel is doing: recognizing significant elements of his environment and communicating with his caregivers.

The semantics row of Figure 5 shows that Nigel is making one kind of meaning, the ideational distinction among the caregivers comprising his meaning-group, but there is no equivalent discrimination of the types of interpersonal doing that is going on contextually.

The expression row of Figure 5 shows the set of three sounds that Nigel pairs with his three meanings. These are all spoken in a high level tone, but differ from each other in terms of articulation. In Nigel's case, unlike that of the Campbell's monkeys, the number of expressions equals the number of meanings.

3.2 Nigel's shift from indexical signs to symbolic signs

Just after beginning to walk, Nigel introduced intonational differentiation into his call system. Instead of a single high level tone, he now had two contrasting tones: falling vs. rising. In this shift, Nigel created a new stratum between meaning and expression, lexicogrammar, in which ideational contrasts are handled by the articulation system, and interpersonal contrasts are handled by the intonation system. Finally, these two different systems are mapped onto each other. Like Figure 4 for the monkeys, Figure 6 shows the bit of grammar that Nigel has slotted in between meaning and expression.

3.2.1   Discussion of Figure 6: Nigel's symbolic system

The semantics row of Figure 6 corresponds to the semantics row of Figure 5. But now Nigel is able to make more than ideational meaning. He can also make interpersonal meaning by greeting or seeking his caregivers. The bit of grammar that he has slotted in between the meaning and expression of his indexical calls enables this semantic increase.

The lexicogrammar row of Figure 6 shows 'word' as entry point for two simultaneous systems: 'ideational-referring', where ideational meanings are expressed by phonetic element sequences, and 'interpersonal-addressing', where interpersonal meanings are expressed by vocalic impulse sequences which change in frequency in time. This grammar thus differs from the Campbell's monkey grammar (Figure 4), in that it systematizes words rather clausette. Nigel recruited intonational sequencing rather than sequencing 'words' in the way the Campbell's monkeys do. The point is not that Nigel has recruited different dimensions of phonetic patterning to make his distinctions. As he moves into adult English he will make choices in many more systems, and will express these through patterns in many units: morphemes in words, words in phrases, phrases in clauses, intonation contour in sequences. But all that is for his future development, as he moves from protolanguage to the adult system. What he has done here is to make

Semantics	Ideational: three types of caregiver (mother, father, and nurse)	Interpersonal: two kinds of address (greeting and seeking)
Lexicogrammar slotted in between semantics and expression curly brackets indicate simultaneous 'and' choices; square brackets indicate alternative 'or' choices	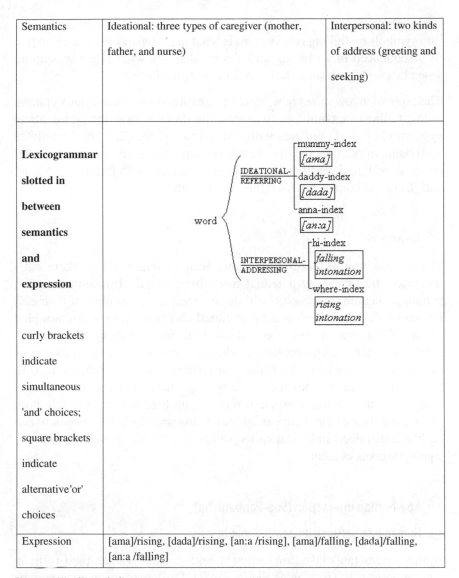	
Expression	[ama]/rising, [dada]/rising, [an:a /rising], [ama]/falling, [dada]/falling, [an:a /falling]	

Figure 6: Nigel's symbolic system

the basic shift from indexical to symbolic meaning by introducing a single additional system which is simultaneous with the first.

The ideational choices mummy-index, daddy-index, anna-index are made simultaneously with the interpersonal choices we are glossing as 'hi'-index or 'where'-index. The two sets of indexical choice are now related in a symbolic system. When making a choice in one set, Nigel must also make a choice in the other. The two indexical systems are necessarily related in the symbolic system, just as they were for the monkeys. Turning indices into symbols by linking two systems is what the monkeys were doing when they sequenced their 'boom' and alarm calls, and what Nigel was doing when he combined word choice and intonation contour.

The expression row shows how Nigel's five expressions ([ama], [dada], [an:a], rising, falling) are combined to express the six choices in the grammatical system: (1) [ama] conflates with rising intonation, (2) [dada] conflates with rising intonation, (3) [an:a] conflates with rising intonation, (4) [ama] conflates with falling intonation, (5) [dada] conflates with falling intonation, and (6) [an:a] conflates with falling intonation.

3.3 Summary

The introduction of word-order choice in the Campbell's monkeys calls increased their meaning potential from three indexical meanings to five symbolic meanings, achieved without increasing the number of their call inventory. The introduction of intonational choice in Nigel's case doubled his meaning potential from three indexical meanings to six symbolic meanings. Nigel increased his inventory: where there was one intonation contour (level), there were now two (falling or rising). The Campbell's monkeys mapped ideational and textual meanings together in a clausette, and Nigel mapped ideational and interpersonal meanings together in a word. In both cases the power of the wedge of grammar to make the difference between highly constrained indexical call systems and infinitely elastic symbolic sign systems is evident.

4   Ape-human interaction (Sue-Panbanisha)

4.1 Human interpersonal discourse semantics

Human interactants take turns when speaking, and organize their turns in a systematic way (Eggins & Slade, 1997). Discourse exchanges consist of at least two turns, and each turn consists of one or more moves. The discourse move system branches into initiating and sustaining moves. A speaker's initiating move (offer, command, statement, question) both sets the agenda for negotiation, and constrains a listener's sustaining move by assigning supporting, or preferred, complementary roles (accept, comply, acknowledge, answer). But sustaining moves are not within the speaker's control, and listeners who in turn become speakers have the option of adopting confronting, or dispreferred, roles (reject, refuse, contradict, disengage). Figure 7 displays these basic discourse moves.

Initiating moves	Sustaining moves	
	Supporting	Confronting
Offer	Accept	Reject
Command	Comply	Refuse
Statement	Acknowledge	Contradict
Question	Answer	Disengage

Figure 7: Basic moves in discourse

4.2 Monkey alarm discourse

Turn-taking exists in a very limited way, since an individual Campbell's or Diana monkey can either emit an alarm call or echo the alarm call of another monkey. A Campbell's monkey alarm call is analogous to an initiating move in human discourse, but since the Diana monkeys' sustaining calls are echoic, there is no question of negotiation. On the other hand, the introduction of the 'boom' call before an alarm call creates something like a preferred (silence) and dispreferred role (vocalization) for the Diana monkeys.

4.3 Nigel's discourse

Nigel's recruitment of intonation to express interpersonal meaning symbolically took place in the context of interaction with caregivers. Nigel's words with composite meaning, e.g. 'where' and 'daddy', would express initiating moves in any discourse with caregivers, and would elicit supporting sustaining moves from them in response.

4.4 Bonobo-human discourse

The ability of bonobos at the Language Research Center, Georgia State University to interpret human lexicogrammar is extensively documented (Savage-Rumbaugh et al.,1993; Savage-Rumbaugh & Lewin, 1994; Savage-Rumbaugh, Shanker & Tayor, 1998). When Kanzi, one of the bonobos at the LRC, vocalizes during interspecies discourse, his vocalizations alternate with human speech in a non-random distribution (Taglialatela et al., 2004; Taglialatela, Savage-Rumbaugh & Baker, 2003). This suggests the non-randomness has a two-part cause: Kanzi's ability to interpret human lexicogrammar gives him the further ability to initiate and sustain a complex discoursal negotiation with a human by the expressive means of a lexigram board, as discussed in Chapter 1.

4.4.1   *A dialogue between Panbanisha and Sue*

The context for the dialogue is that Panbanisha has just finished playing an electronic piano in collaboration with a group of professional musicians. At the end of her session, she pressed the lexigram GOOD. Kanzi's turn to play is next. He is out of sight in an adjacent room, but near enough to be able to overhear the dialogue between Panbanisha and Sue. A transcript of this dialogue, divided into discourse moves, can be found in the appendix. The first 21 moves of the discourse exchange revolve around a discussion between Panbanisha and Sue about Kanzi's willingness to cooperate. In moves 22 and 23, Panbanisha and Sue jointly ask Kanzi himself about his willingness to cooperate.

As her means of expression in this particular dialogue, Panbanisha does not use lexigrams; rather she produces two types of vocalizations: proximal and distal. In proximal vocalizations the interactants are close enough for eye contact; and in distal vocalizations the interactants are farther away from each other. When Panbanisha addresses Sue, her proximal vocalizations resemble the 'peeps' described by de Waal (1988), though there are differences. For example, de Waal's food peeps have an average peak frequency of 2.3 kHz. in the fundamental, and de Waal's alarm peeps have an average peak frequency of 1.7 kHz. The highest of Panbanisha's proximal vocalizations reaches a peak of 1.891 kHz., and the lowest reaches a peak of 1.375 kHz. Similarly, Panbanisha's distal vocalizations resemble de Waal's hoots, but none of them reach an average peak frequency of 2.3 kHz., with the highest reaching a peak of 1.974 kHz., and the lowest a peak of 1.675 kHz. Panbanisha's proximal vocalizations are also more varied than de Waal's peeps. Panbanisha's proximal vocalizations show four different intonational contours: level, falling, rising, and rising-falling.

Human discourse is a joint, cooperative activity in which the interactants share semiotic space. In discourse, humans choose from a well-established system of move sequences. Humans interacting together in Panbanisha's presence do exactly this, and Sue and other humans interacting with Panbanisha are also operating the system of move sequences. Panbanisha shows evidence of choosing within this system too, because her vocalizations occur where turns and moves would be made if Sue had been interacting with a human. Sue, for her part, interprets and responds to Panbanisha's vocalizations as systematic turns and moves. In other words, Sue and Panbanisha are jointly creating and maintaining a semiotic environment of symbolic negotiation in which Sue uses human language, and Panbanisha uses vocalizations. A strong bit of evidence for this claim is that it is the discourse context itself that causes Panbanisha to change from talking *about* Kanzi to talking *to*

Kanzi. Figure 8 shows the placement of Panbanisha's vocalizations in terms of moves in human discourse.

Initiating moves	Sustaining moves	
	Supporting	Confronting
Sue: (1) question	Pan: (2) maintain: voc:prox 1, voc:prox 2 (overlap)	
	Pan: (2) answer: voc:prox 3, voc:prox 4	
Sue: (3) question	Pan: (4) maintain: voc:prox 5 (overlap)	
	Pan: (4) answer: voc:prox 6, voc:prox 7	
Sue: (5) question		Pan: (6) disengage: silent turn
Sue: (7) question		Pan: (8) disengage: silent turn
Sue: (9) statement	Pan: (10) maintain: voc:prox 8 (overlap)	
	Sue: (11) maintain	
silence		
Sue: (12) question		Pan: (13) disengage: silent turn
Sue: (14) question		Pan: (15) disengage: silent turn
Peter: (16) statement	Sue: (17) acknowledge	
Sue: (18) statement	Pan: (19) maintain?: voc: prox 9	Pan: (19) contradict?: voc: prox 9
Sue: (18) statement	Pan: (19) maintain?: voc: prox 10	Pan: (19) contradict?: voc: prox 10
Sue: (20) offer	Pan: (21) accept: voc:prox 11	
Sue: (22) question (overlap)	Pan: (23) question? Voc:dist 12, 13, 14, 15, 16, 17, 18, 19	

Figure 8: Placement in turns in human discourse

4.4.2 Discussion of Figure 8

In Figure 8, the columns are organized in terms of discourse exchanges: the first column contains initiating moves, the second two columns contain two kinds of sustaining moves: supporting and confronting. There are sixteen initiating moves in the rows which make up this conversation. In Figure 8, the speaker is identified, numerals indicate turns, and turn numbers are followed by the type of move. For example, 'Sue: (1) question' means that Sue is the speaker, that this is the first turn in the exchange, and that the type of initiating move is question; 'Pan: (2) maintain… (overlap)' means that Panbanisha is the speaker, that this is the second turn in the exchange, that the type of sustaining: supporting move is maintain, and that it overlaps with Sue's speech.

We interpret Panbanisha's vocalizations in the second and third columns of the first row as actions made within the human discourse pattern because they occur at appropriate points, and because Sue interprets them as moves. For the same reason, we interpret Panbanisha's silences as moves, but only when they occur at appropriate points, and when Sue responds to them as such. Where Panbanisha's vocalizations overlap Sue's turns, we interpret them as maintaining moves, in the same way that a nod made in the course of someone's statement or question is a maintaining move encouraging the speaker to keep going.

Proximal vocalizations: turns (1)–(4)

In (1), Sue asks the question 'Do you think Kanzi will like (voc:prox 1) to play (voc:prox 2)?' In the course of this, Panbanisha cooperates by her overlapping (2) (voc:prox 1 and voc:prox 2), judged to be maintain, because of Sue's response (she keeps going, rather than treating either of the vocalizations as an interruption to which she has to respond). After Sue has finished the question, Panbanisha vocalizes twice more in (2) (voc:prox 3 and voc:prox 4), judged to be answer, because of Sue's response in (3), 'Do you think he'll do a good (voc:prox 5) job?' a question which implies that Sue has interpreted Panbanisha as having given a positive answer to her first polar question, since in order to do a good job, Kanzi will have to play the piano. Unlike a lexical information-seeking question which requires a lexical answer, a polar question requires only a yes or no answer. Whether or not Sue has been correct in interpreting Panbanisha's third and fourth vocalizations as indicating positive polarity, they constitute supporting moves because they occur where an answer needs to be.

This pattern is repeated in (3), with Sue asking the question 'Do you think he'll do a good (voc:prox 5) job?'. Panbanisha's overlapping (4) maintain (voc:prox 5) is followed, after Sue has finished her question, by (4) answer (voc:prox 6 and voc:7).

Proximal vocalizations: turns (5)–(15)

Sue's next move, (5) question 'Will he be nice when he's in here?', elicits a completely different response from Panbanisha – a silent turn (6) disengage. Even though Panbanisha is not vocalizing, it is still a turn in discourse because Sue interprets it as such, since her next move (7) question asks 'Not so sure about being nice hunh?', spoken with rising intonation. Panbanisha's (6) disengage is a confronting rather than a supporting move, since she is opting out of the conversation altogether where an answer is required.

Panbanisha's (8) disengage also avoids answering Sue's (7) question. In addition to indicating that she is interpreting Panbanisha's (8) disengage as a turn in discourse, Sue's (9) statement 'Aw, Sue told Kanzi he had to be nice (voc:prox 8)' indicates that she is also interpreting Panbanisha's silent turn (8) as a negative response to her (7) polar question. Panbanisha's (10) maintain (voc: prox 8) overlaps with Sue's (9) statement, and Sue responds to it, with a supporting move, (11) maintain 'Yeah, Sue told him'.

At this point in the dialogue there is a period of silence. This silence, however, is not a turn in discourse, because the move system doesn't require anything of Panbanisha, and indeed Sue doesn't interpret Panbanisha's silence here as playing a role in the ongoing discourse.

In (12) question, 'Do you think he'll listen to me?', Sue asks a different but related question, which elicits another silent turn, (13) disengage, from Panbanisha. Sue follows up with (14) question, 'You're not so sure he'll listen to me?' (rising intonation). This is similar to her previous (7) question, 'Not so sure about being nice hunh?'. This in turn elicits yet another silent turn from Panbanisha, (15) disengage.

Proximal vocalizations: turns (16)–(21)

At this point, the musician Peter Gabriel joins the conversation with (16) statement, 'He's gotta be nice if we're going to make music together', to which Sue responds with (17) acknowledge, 'Yeah'.

Sue's (18) statement, 'Well, if Kanzi's not nice may(voc:prox 9) maybe I should(voc:prox 10) tell him to stay there', is a departure from her pattern of questioning in (1), (3), (5), (7), (9), (12), and (14). Grammatically, the statement consists of two clauses: the first clause expresses a condition, and the second clause is a highly modalized expression of consequence, and amounts to a threat: Kanzi may not get to play the piano.

Panbanisha's overlapping vocalizations, (19) sustain (voc:prox 9 and voc: prox 10), could fit into the previously establish pattern of maintain. It is also possible, however, that Panbanisha is sustaining the discourse in a confrontational way. This would be in accord with the way Sue's two previous questions (9) 'Will he be nice when he's in here?' and (11) 'Not so sure about being nice hunh?' elicited silent turns. That is, by vocalizing, as opposed to opting out of the discussion through silence, Panbanisha is contradicting rather than acknowledging Sue's (18) statement.

Sue now tries a new approach, with (20) offer (more precisely a suggestion), 'Shall we ask him if he's going to be good?'. Although the grammar is interrogative mood, Sue is not asking for polar information; what she wants is acceptance of her suggestion that she and Panbanisha perform a joint action: asking Kanzi directly. This is quite different from speculating about his likely behaviour. Previously, the agenda for negotiation consisted of a series of demands for information. Here, however, the agenda has changed. Sue is not demanding information from Panbanisha, she is offering to join with Panbanisha in a new course of action.

Panbanisha's (21) accept (voc:prox 11) is judged to constitute accept, because she does indeed perform her part of the joint action by means of a distal vocalization immediately thereafter in (23) initiating. This overlaps Sue's (22) question, 'Kanzi, are you going to (voc:dist 12) be good when (voc:dist 13) we come in (voc:dist 14) when you come (voc:dist 15) in (voc: dist 16) here? (voc:dist 17 voc:dist 18 voc:dist 19)'.

Distal vocalizations: turns (22)–(23)

What has caused Panbanisha to emit these distal vocalizations after a run of eleven proximal vocalizations? Although it might be thought that Panbanisha was automatically reacting to the key-word 'Kanzi', the evidence does not support this. On the three previous occasions when Sue has mentioned Kanzi by name (1), (9), and (18), Panbanisha responded with proximal vocalizations, not distal vocalizations. That is, Panbanisha has very different reactions to the word 'Kanzi' depending on where it occurs in the system of discourse moves.

We have analyzed Sue's interpretations in some detail, but have had little to say about Panbanisha's interpretations, except at the discourse-semantic level of moves. We can, however, draw inferences about Panbanisha's lexicogrammatical interpretations, since in order to respond to the clauses that constitutes Sue's moves in a non-random way, she must to some degree be able to process those clauses.

Like all casual conversation, this dialogue is heavily pronominal. There are three occurrences of 'I/me', two occurrences of 'we', four occurrences of 'you', and eleven occurrences of 'he/him'. All of these pronouns refer not only to named individuals, but also to grammatical participant roles in different kinds of clausal processes, such as thinking, doing, being and saying. For example, Sue assigns 'Kanzi' the participant role of Actor in a doing process in (3), 'Do you think he'll do a good(voc:prox 5) job?', where 'a good job' is the Goal of his doing; for another example, Sue assigns 'he' the participant role of Carrier of the Attribute 'nice' in a being process in (5), 'Will he be nice when he's in here?'. 'Kanzi' is referred to either by proper name or pronoun in every single one of Sue's and Peter Gabriel's clauses in this dialogue ('Kanzi/him' is interpretable by ellipsis in (7)).

The placement of Panbanisha's distal vocalizations implies that she has been tracking pronominally referenced participant roles throughout the dialogue. There are three such pronouns in Sue's (20) suggestion, 'Shall *we* ask *him* if *he*'s going to be good?'. 'We' refers to Sue and Panbanisha, and 'him' and 'he' refer to Kanzi. But the pronouns also refer to participants in processes: 'we' are co-sayer participants in a process of saying, 'him' is the target participant in the process of saying, and 'he' is the Carrier participant in the process of being, where 'good' is the Attribute ascribed to him. Panbanisha thus *acts out* what was coded in the grammar of Sue's suggestion, when in (22) and (23), 'we' (Sue and Panbanisha) jointly address 'him' (Kanzi), using human speech and distal vocalizations respectively. Without responding to this grammatical structure, Panbanisha could not have played the meaningful role in the discourse that she does.

4.5 The wedge of discourse

In Sections 2 (monkey calls) and 3 (Nigel), we discussed how the wedge of grammar between semantics and expression opened up a potential for exponential growth in the ability to make meanings. In this section, we have seen that discourse, building on the increase in Panbanisha's ability to interpret meaning made possible by the wedge of grammar, becomes a wedge between semantics and context. That is to say, when Panbanisha accepts and then enacts Sue's suggestion, *Sue's discourse itself has become part of relevant context for Panbanisha.* Since discourse puts pressure on lexicogrammar to differentiate, in order to enable ever more complex negotiations in a co-evolutionary dynamic, it too is a strong candidate for an evolutionary prerequisite for language.

5   Conclusion

The Diana and Campbell's monkey 'snapshot' reveals the wedge of grammar, an ability to differentiate and then combine different kinds of indexical signing in one utterance. This bit of grammar changed the fixed discoursal pattern of call and automatic response to a meaningful choice between echoing and not echoing. It is a first step, but only the first step, in a co-evolutionary process: a grammatical development of this sort would make possible more complex discourse, and more complex discourse in turn would put pressure on grammar to develop yet more complexity.

With the Nigel 'snapshot' we see a parallel between the co-evolutionary processes of deep time and the adult-child dyad in developmental time. Nigel's web of caregivers is complex: adult humans who come and go, who have different relationships with him, and who provide goods and services of many types. There is more going on than he can make meaning about with a simple call system. When Nigel recruits intonation in addition to articulation, this is also a first step making possible more complex discourse in response to the cultural pressure of the adult symbolic language to which he is exposed. With this step he begins to make meaning about the world symbolically, and he is poised to use the symbolic language system of his caregivers.

In the Panbanisha 'snapshot', by addressing Kanzi, Panbanisha is not only responding to the role that Sue's language is playing in context, but also giving her distal vocalization a role to play in this context. That is, not only is Panbanisha able to mean more, but also she is able to do more; and with this discourse ability, Panbanisha has increased her range of doing far beyond 'saying', since the wedge of discourse enables languaged bonobos to leave a world of passive happenings and enter into an expanding world of cultural doings. Panbanisha's, as well as Kanzi's, music-making is a doing which is embedded in an on-going discourse context. As such, it is like their tool-making (Schick et al., 1999; Savage-Rumbaugh, Fields, and Taglialatela, 2001), which is also embedded in an on-going discourse context. Cultural creativity of this kind is far beyond the capabilities of unlanguaged bonobos at the LRC.

Even a very small amount of symbolic language and discourse makes a great difference, and our hominid ancestors, with the wedge of discourse, building upon the wedge grammar, would have similarly increased the range of their cultural doing. This in turn would have put pressure upon their emerging lexicogrammatical and discoursal abilities in a robust co-evolutionary dynamic.

References

Benson, J., Fries, P., Greaves, W., Iwamoto, K., Savage-Rumbaugh, S. and Taglialatela, J. (2002) Confrontation and support in bonobo-human discourse. *Functions of Language*. 9(1): 1–38.

Cheney, D. L. and Seyfarth, R. M. (1990) *How Monkeys See the World: inside the mind of another species*. Chicago: University of Chicago Press.

Deacon, T. W. (1997) *The Symbolic Species: the co-evolution of language and the brain*. London and New York: W. W. Norton.

de Waal, F. (1988) The communicative repertoire of captive bonobos (Pan paniscus), compared to that of chimpanzees. *Behaviour* 106: 183–251.

Eggins, S. and Slade, D. (1997) *Analysing Casual Conversation*. London and Washington: Cassell.

Fawcett, R., Tucker, G. and Lin, Y. (1993) How a systemic functional grammar works: the role of realization in realization. In H. Horacek and M. Zock (eds) *New Concepts in Language Generation: planning, realization and system*. London and New York: Pinter.

Halliday, M. A. K. (1995) On language in relation to the evolution of human consciousness. In S. Allén (ed.) *Of Thoughts and Words: proceedings of the Nobel symposium 92: the relation between language and mind*. 45–84. Singapore: Imperial College Press.

Halliday, M. A. K. (1994) *An Introduction to Functional Grammar*. London: Edward Arnold.

Halliday, M. A. K. (1975) *Learning How to Mean: explorations in the development of language*. London: Edward Arnold.

Halliday, M. A. K. and Matthiessen, C. M. I. M. (1999) *Construing Experience Through Meaning: a language-based approach to cognition*. London and New York: Cassell.

Martin, J. R. (1992) *English Text: system and structure*. Philadlephia/Amsterdam: John Benjamins.

Matthiessen, C. M. I. M. (1995) *Lexicogrammatical Cartography: English systems*. Tokyo: International Language Sciences Publishers.

Savage-Rumbaugh, S., Shanker, S. and Taylor, T. (1998) *Apes, Language, and the Human Mind*. New York: Oxford University Press.

Savage-Rumbaugh, E. S., Fields, W. M. and Taglialatela, J. P. (2001) Language, speech, tools, and writing: a cultural imperative. *Journal of Consciousness Studies* 8(5–7): 273–92.

Savage-Rumbaugh, S. and Lewin, R. (1994) *Kanzi: the ape at the brink of the human mind*. New York: John Wiley Publishers.

Savage-Rumbaugh, S., Murphy, J., Sevcik, R., Brakke, K., Williams, S. and Rumbaugh, D. (1993) *Language Comprehension in Ape and Child*. Monographs of the Society for Research in Child Development. Chicago: University of Chicago Press.

Schick, K. D., Toth, N., Garufi, G., Savage-Rumbaugh, S., Rumbaugh, D. M. and Sevcik, R. A. (1999) Continuing investigations into the stone tool-making and tool-using capabilities of a bonobo (Pan paniscus). *Journal of Archaeological Science* 26: 821–32.

Taglialatela, J., Savage-Rumbaugh, S. and Baker, L. (2003) Vocal production by a language-competent Pan paniscus. *International Journal of Primatology* 24(1): 1–17.

Taglialatela, J., Savage-Rumbaugh, S., Rumbaugh, D. M., Benson, J. and Greaves, W. (2004) Language, apes and meaning-making. In G. Williams and A. Lukin (eds) *The Development of Language: functional perspectives on species and individuals*. London: The Continuum International Publishing Group.

Wu, C. (2000) *Modelling Linguistic Resources: a systemic functional approach*. Macquarie University: Linguistic PhD thesis.

Zuberbühler, K. (2005) Linguistic prerequisites in the primate lineage. In M. Tallerman (ed) *Language Origins: perspectives on evolution*. Oxford: Oxford University Press.

Zuberbühler, K. (2002) A syntactic rule in forest monkey communication. *Animal Behaviour* 63: 1–7.

Appendix 1

Transcript of dialogue divided into discourse moves

Key: Parenthesis = Panbanisha's vocalizations overlapping with Sue's speech
Numerals = Turns

(1) Sue to Pan: Do you think Kanzi will like(voc:prox 1) to play(voc: prox 2)?

(2) Pan to Sue: voc:prox 1,voc:prox 2, voc:prox 3, voc:prox 4

(3) Sue to Pan: Do you think he'll do a good(voc:prox 5) job?

(4) Pan to Sue: voc:prox 5, voc:prox 6, voc:prox 7

(5) Sue to Pan: Will he be nice when he's in here?

(6) Pan to Sue: silence 1

(7) Sue to Pan: Not so sure about being nice hunh?

(8) Pan to Sue : silence 2

(9) Sue to Pan: Aw, Sue told Kanzi he had to be nice (voc:prox 8)

(10) Pan to Sue: voc:prox 8

(11) Sue to Pan: Yeah, Sue told him

(12) Sue to Pan: Do you think he'll listen to me?

(13) Pan to Sue: silence 3

(14) Sue to Pan: You're not so sure he'll listen to me?

(15) Pan to Sue: silence 4

(16) Peter G (to whole group?): He's gotta be nice if we're going to make
    music together

(17) Sue to whole group: Yeah

(18) Sue to Pan: Well, if Kanzi's not nice may(voc:prox 9)

(18) Sue to Pan: maybe I should(voc:prox 10) tell him to stay there

(19) Pan to Sue: voc:prox 9

(19) Pan to Sue: voc:prox 10

(20) Sue to Pan: Shall we ask him if he's going to be good?

(21) Pan to Sue: voc:prox 11

(22) Sue to Kanzi: Kanzi, are you going to(voc:dist 12) be good when (voc:
    dist 13) we come in (voc:dist 14) when you come(voc:dist 15)
    in(voc:dist 16) here?(voc:dist 17 voc:dist 18 voc:dist 19)

(23) Pan to Kanzi: voc:dist 12, voc:dist 13, voc:dist 14, voc:dist 15, voc:dist
    16, voc:dist 17, voc:dist 18, voc:dist 19

4   The multistratal dimension: a methodology for phonemic analysis of vocalizations of language competent bonobos

James D. Benson, Meena Debashish, William S. Greaves, Jennifer Lukas, Sue Savage-Rumbaugh and Jared Taglialatela

Abstract

Many different things happen at the same time in human language. In the science of linguistics these are traditionally treated in analytically separate sub-disciplines, for example phonetics, phonology, morphology, lexical semantics, and syntax. Our method for examining the vocalizations of language competent bonobos is more integrated; the tradition of Systemic Functional Linguistics regards language as a whole which must be analyzed as relations between the different systems in five strata: context, semantics, lexicogrammar, phonology and phonetics. Because the essence of this approach is the relationships, the wholeness of language is emphasized, but at the same time the systems discovered in each stratum display the discriminations which make language's creation of information possible. Language as a whole offers literally billions of choices – a seemingly infinite chaos. This chapter focuses on the process of discrimination through choices within systems at all strata, but in particular those systems related to sound which make it possible for an English speaking human interpreter to recognize English words in the distinctly non-human sounds emitted by a language competent bonobo engaged in discourse with the human.

1   Introduction

In Chapter 1 we focused on human discourse patterns of negotiation, presenting evidence for Kanzi's ability to integrate lexigrams as a medium of expression for his interactions with Sue Savage-Rumbaugh in a negotiation based on Kanzi's original command: 'GO OPEN GROUPROOM'.

In Chapter 2 we presented evidence for Kanzi's ability to process human symbolic lexicogrammar. The experiment furnishing the data consisted of novel sentences that were designed to test Kanzi's syntactic abilities. In terms of discourse, the experiment consists of a series of reiterated English commands spoken by the experimenter, which positioned Kanzi in the role of compliance. In the 'GO OPEN GROUPROOM' dialogue, Kanzi took on the commanding role and Sue took on the complying role, although she did not, in fact, comply. These discourse roles were reversed in the experimental dialogue, where Kanzi took on the complying role, and in fact did his best to comply with the various novel commands.

In Chapter 3 we presented evidence for Panbanisha's ability to integrate her vocalizations as a medium of expression into human discourse patterns of negotiation. Just as was the case for Kanzi's use of lexigrams, Panbanisha's vocalizations occurred where moves in the system of discourse were expected. The vocalizations were of two types, proximal and distal, and Panbanisha seemed to direct her proximal vocalizations to Sue, and her distal vocalizations to Kanzi (who was not in her purview).

In this chapter we look closely at Kanzi's proximal vocalizations, and their interpretation by humans as words of spoken English, for the following reasons:

- Kanzi's behaviour in conversations with Sue resembles human discourse in so far as they take turns where appropriate, as discussed in Chapter 1;

- Kanzi uses his vocalizations to communicate during linguistically mediated interactions with humans (Taglialatela et. al., 2004);

- Kanzi is capable of producing unique, non-species typical bonobo sounds (Hopkins and Savage-Rumbaugh, 1991);

- Kanzi is able to modulate his vocal output by selectively producing certain sounds in certain semantic contexts (Taglialatela, Savage-Rumbaugh and Baker, 2003);

- Kanzi is able to change the shape of his vocal tract
  (Savage-Rumbaugh, Fields and Spircu, 2004).

It is thus reasonable to treat Kanzi's vocalizations as approximations of English words (which researchers and caregivers who work with him do on a daily basis). We propose a methodology designed to enable the study of how humans can make such interpretations, and show that when one takes this approach, observable (and quantifiable) differences are readily apparent in the structure of Kanzi's vocalizations.

2   Methodological assumptions: the multilayered informativeness of systemic choice

Each of the strata (context, semantics, lexicogrammar, phonology or phonetics) contributes a different kind of information. The combination of these different kinds of information is what makes possible the human interpretation of bonobo vocalizations and the bonobo interpretation of human vocalizations.

2.1 Integration and differentiation: two fundamental properties shared by consciousness, distributed neural systems, and language

In Chapter 2 we provided evidence for Kanzi having some degree of higher-order consciousness.

Integration and differentiation play similar roles in three distinct but interdependent phenomena: consciousness, neuronal organization and language.

Edelman and Tononi (2000: 111) posit a high degree of both integration and differentiation as the two fundamental properties of *consciousness*:

> first, consciousness is highly integrated or unified – every conscious state constitutes a unified whole that cannot effectively be subdivided into independent components – and second, at the same time, it is highly differentiated or informative – there is an enormous number of different conscious states, each of which can lead to different behavioural consequences.

Edelman and Tononi also posit a high degree of both integration and differentiation as the two fundamental properties of *neuronal organization*:

> the distributed neural processes underlying consciousness also share these properties: They are highly integrated and, at the same time, highly differentiated. We believe that this convergence between neurobiology and phenomenology is not mere coincidence.

The nearly infinite integrated and differentiated complexity of neural proc-esses is matched by the nearly infinite integrated and differentiated com-plexity of the *language system*. Since language is one of the tools the brain uses in creating higher-order consciousness (Edelman and Tononi (2000: 193-99), it is not surprising that language in its wholeness and diversity shares the properties of a high degree of integration and differentiation with that consciousness and with the 'distributed neural processes underlying consciousness'. This is not to say, of course, that language systems and neuronal systems are isomorphic. They are not. But they are intimately related.

Edelman and Tononi (2000: 137) see external stimuli as significant in the formation of any individual animal's neuronal systems:

> What must be determined theoretically and measured, then, is how the intrinsic dynamic relationships among specialized neuronal groups in an adult brain become adaptively related, over time, to the statistical structure of the environment – the average, over time, of all signals characteristic of the environment received by an animal. Moreover, given that at any given time, most neuronal groups in the brain are predominantly affected by information from other parts of the brain (or intrinsic information), the moment-to-moment contribution of the information that is actually provided by the environment (extrinsic information) must also be determined.

Language is an extraordinarily complex system in our environment and plays an important role in shaping the neuronal systems which support higher-order consciousness by providing the stimulus of human speech sounds rich in complex information to neural systems.

Differentiating between a large number of *conscious states*, supported by *neural processes*, 'constitutes information, which is the reduction of uncertainty among a number of alternatives.' The information generated may lead to 'different behavioural outputs' (Edelman and Tononi, 2000: 126). *Language*, too, yields information through differentiation, since language consists of thousands of systems, each of which presents choices which are related by probability to choices made in other systems. The reduction of uncertainty achieved by any particular choice will lead to different linguistic outputs.

2.2 The MOVE system: an example of differentiation and integration internal to the semantic stratum

A simplified version of the differentiations made at the semantic stratum by the interpersonal MOVE system (discussed in Chapter 1, Sections 3, 4.1; Chapter 3, Section 4) is shown in Figure 1.

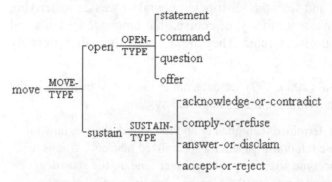

Figure 1: Simplified MOVE system

Looking around within the semantic stratum, where the MOVE system is located, we see the organization of different meanings humans have developed to exchange information and goods and services as they go about the business of interacting as social groups.

Meanings are organized taxonomically, so that any choice in the system both integrates and differentiates. For example, an [open] move integrates [statement] [command] [question] and [offer], and differentiates these from [sustain]. The alternative choice, [sustain], integrates [acknowledge or contradict], [comply or refuse], [answer or disclaim], or [accept or reject] – differentiating these from [open].

The system thus constrains the way a speaker must differentiate: alternatives can only be selected within the parameters of the system, and any particular selection (e.g. [open]) rules out all other choices in the system (e.g. [sustain] and everything integrated by [sustain]). Differentiations integrated by the selection of [open] ([statement], [question], [command] and [offer]) remain available. This is what makes any choice within the system highly informative.

2.3 The MOVE system: looking upward from semantics to context and downward from semantics to lexicogrammar

The MOVE system enables the creation of one type of information – interpersonal semantic information. Many different types of information are created at the various strata: *behaving information* by systems in context, *meaning information* by systems in semantics, *wording information* by systems in lexicogrammar and *expressing information* by systems in phonology or graphology.

Selections of information in the different strata are related through probability. If, for example, we look up from the semantic stratum to the context stratum, we see that the different choices in the MOVE system correlate with differentiated social behaviours in context, which are shown in upper case letters in the boxes of Figure 2.

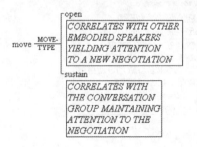

Figure 2: Initial MOVE types correlated with differentiated contexts

Figure 3 shows the further differentiations of [open] (cf. Figure 1). Like Figure 2, Figure 3 looks upward. Differentiated social behaviours of the context stratum (represented in upper case) are correlated with the differentiations in the MOVE system. Unlike Figure 2, Figure 3 also looks downward to the lexicogrammatical stratum. Differentiations in the MOVE system correlated with choices in the lexicogrammatical MOOD system (represented by lower-case italics).

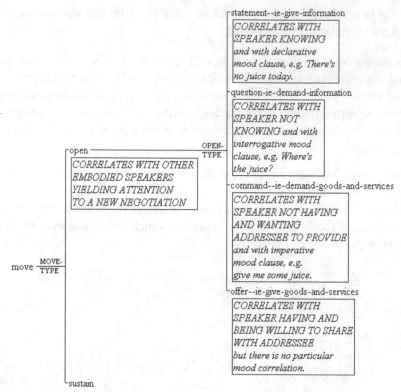

Figure 3: MOVE types with [open] differentiation

Looking upwards from meaning to behaving, the context of situation, we see that the differentiations [statement], [question], [command] and [offer] generalized as [open] each have different correlations with the speaker's relationship to the others in the communication group. These are shown in upper case. But there is also a correlation with wording. Looking downward from meaning to wording (shown in lower case), we see different types of probable clause mood for [statement], [question] and [command]. In the case of [offer], however, the mood choices seem equally probable: [declarative] 'There's juice in the fridge', [interrogative] 'Would you like a juice?', [imperative] 'Help yourself to a juice!' all serve to indicate a speaker's meaning, *i.e.* 'I will give goods and services'.

2.4 The informativeness and behavioural outcome of choices in the MOVE system

The simple system shown in Figure 1 thus shapes the delivery of a considerable range of information. For example, a speaker making a move must choose between [open] and [sustain], and by choosing [open] informs the addressee, by that selection in the system, of the beginning of a negotiation. This information has the behavioural consequence that the addressee is under pressure to respond. But the nature of the response depends on a further discrimination made in the [open] move, within the four choices offered by the system. Each of these choices is informative in a different way, and each, therefore, leads to different language or other behavioural outcomes on the part of the addressee. If the opening move is [statement], the expected response is [acknowledge]. If it is [command], the expected response is [comply]. If it is [question], the expected response is [answer], and if it is [offer], the expected response is [accept]. Each of these responses is differentially informative, precisely in terms of the alternatives excluded. An [acknowledge], for example, 'fits' a [statement] because it is not [comply], [answer] or [accept]. Choice of a dispreferred alternative, such as [contradict] rather than [acknowledge] as a response to [statement] is, of course, equally informative. The response of either [acknowledge] or [contradict] would, of course, become a stimulus in its own right. Whatever the choice, it is the system which organizes the exchange of information or of goods and services in a dialogue.

2.5 Differentiation in phonology

Just as language systems guide and control both speaker and hearer differentiation in semantics (e.g. the MOVE system) and in lexicogrammar (e.g. the MOOD system), differentiation is guided and controlled in phonology through choices offered by the PHONEME system. Acts of discrimination in the different strata are related to each other by correlation. Given a particular contextual setting (e.g. feeding), a particular discourse path (e.g. [offer]), and a particular MOOD (e.g. [question: '(would you like a) xxx?']), there is a probability that a speaker will be choosing from a small set of words, e.g. *juice* or *onion*. Just as an interpreter must discriminate the MOVE, MOOD and LEXICAL WORD choices, he or she must discriminate the sounds. This interpretive process can be thought of as hypothesizing and confirming: a particular string of phonemes distributed in syllables is probable – are these recognized as icons in the English sound system (following the principles of iconic interpretation discussed in Chapter 2, Section 4)?

The extent of the rich and varied information generated at the strata of context, semantics, and lexicogrammar discussed in Sections 2.1–2.4 means that the interpretation of the phonemic structure of Kanzi's vocalizations is only a part of a much larger interpretive process. It is this larger interpretive process which makes it possible for a hearer to make iconic interpretations (Deacon 1997), that is, to ignore the differences between Kanzi's sounds and human sounds, and recognize that they represent choices made in the same system. Kanzi makes both one syllable and two syllable vocalizations. Therefore, reduction of uncertainty by ruling out all one syllable words is tremendously informative in the process of deciding between *juice* and *onion*, for example, when Kanzi's vocalization could represent either of these in the unfolding context.

Our methodology focuses on the way in which information created by choices made in all strata correlates with the production and interpretation of Kanzi's proximal vocalizations. We are proposing a method for examining the proposition that, although they sound remarkably unlike what we hear coming from the lips of the humans we normally engage in discourse, some bonobo vocalizations are uttered as, and are interpretable as, allophones of English phonemes. That is, different classes which these allophones form are distributed functionally and are allophones of English phonemes – strange and exotic allophones, but they are sounds which contrast with each other and make up a dialect of English which is recognizable by an interpreter who has tuned his or her ear to its nature. Furthermore, because they are classifiable phonetically as vowels and consonants, these sounds are capable of forming syllables. The notion of bonobo allophones may seem far-fetched, yet the ears of interpreters can readily be tuned to the 'unnatural' combinations of sinewaves, as summarized on the Haskins Lab SineWave Synthesis website (http://www.haskins.yale.edu/Haskins/MISC/sws/sws. html). It is the English phoneme system that allows the synthesized sinusoidal utterance to be interpreted as a succession of English phonemes and syllables, even though individual allophones lack the acoustical properties of consonants (noise) and vowels (broadband formants).

Our methodology allows us to ask precise questions about vowel and consonant classification in the bonobo dialect of English, when applied to a large body of spontaneous bonobo-human discourse: Does Kanzi recognize the different classes of vowels in the English phoneme system? If so, can he produce them? That is, does he have his own allophones of these phonemes? Does Kanzi have onset and coda consonants? If so, does he have different classes of consonant, for example /k/ and /g/? Does he have all of the English consonant phonemes, or just some of them? Does he have a recognition command of all the consonant phonemes? If so, does he have his own allophones of these?

3   Methodology

We establish target syllables and phonemes through the process of multi-stratal analysis, using context and discourse structure to discover probable grammatical words. These words identify the target phonemes, which we then compare with the phonetic elements in Kanzi's vocalization of the proposed word.

3.1 Data management: Systemic Coder and Praat

We use Systemic Coder (http://www.wagsoft.com/Coder/index.html) to represent the choices available in the various systems at the different strata. Figure 1 was an example of a simplified system of interpersonal choices in the semantic stratum. Figure 4 shows a fragment of a much fuller version of the same system.

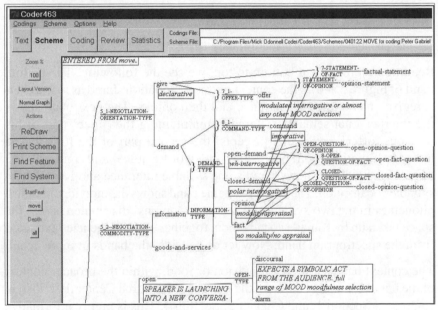

Figure 4: Fragment of the MOVE system in interpersonal semantics

We use the Praat software (http://www.fon.hum.uva.nl/praat/download_win.html) to align analysis at each stratum with the soundwave containing the bonobo-human conversations. Figure 5 shows what this looks like for a short stretch of sound.

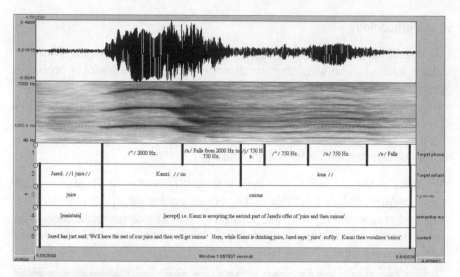

Figure 5: One second of sound with five analysis tiers

3.2 The phonemes of *onion* in multistratal analysis

Reading Figure 5 from bottom to top, we see the following. The bottom band of Figure 5 shows the context of this negotiation: Jared as researcher-caregiver, has offered Kanzi juice and then onions. The next band shows the interpersonal semantics: Jared is maintaining the 'juice' part of his previous [offer], and Kanzi [accepts] the second part of the [offer]. The lexicogrammatical word band shows the words: 'juice' and 'onions'. The target syllable band shows Jared's one syllable utterance and Kanzi's two syllable vocalization. The target phoneme band shows the individual English phonemes in the two syllables of the word 'onions' distributed across the space taken up by Kanzi's vocalization, together with frequencies extracted from the spectrogram band. Now let us consider the bands in more detail.

The context band shows the provision of food, within the broader context of the bonobo-human culture of the Language Research Center described in Chapter 1, Section 1, and Chapter 3, Section 4. This is highly informative because it rules out all other contexts Kanzi is familiar with. In addition, the provision of food is not just a mechanical, non-social act. Jared is providing companionship as well as sustenance. With his hands he is providing food, but at the same time with his mouth he is dialoguing, just as Kanzi is using his hands to drink the juice, but his mouth, as well as drinking the juice is vocalizing at Jared. The two individuals are interacting. Jared is talking to Kanzi about the timing and sequence of what he is doing, and Kanzi is very much engaged with Jared's talk.

This is the type of behaviour which correlates with semantic patterning, and, indeed, the semantic band shows how this interaction is shaped in the way in which English speaking humans typically negotiate: a series of turns which relate to each other in a patterned way. Jared's [offer] is highly informative because it rules out all other opening moves in a negotiation, and predicts Kanzi's choice of [accept] or [reject]. This is informative because any other meaning is very unlikely.

Making and accepting offers is part of interpersonal semantics. In this case, two different kinds of food are being offered. Ideational semantics differentiates juice and onions from all other foods. This too is highly informative, as countless other possibilities are discriminated out.

This ideational meaning correlates with lexicogrammar. The lexicogrammatical band shows that we have found that the word 'onion' is more probable than the word 'juice' – mainly because of the phonology band. We hear two syllables in Kanzi's vocalization, and the spectrogram and waveform bands show why: there are two prominent vowel areas which are clearly separated by a sharp drop in frequency (2000 Hz. to 750 Hz.). In other words, Kanzi's syllabification is highly informative and rules out all one syllable words including 'juice'.

The question still arises: are these syllables English? Could they not be normal bonobo sounds? Here, too, the spectrogram is helpful. de Waal (1988) has provided spectrograms for the vocalizations of bonobos in a zoo setting who have not been socialized in an English speaking culture in the way Kanzi has. Kanzi's vocalization here bears no resemblance to any of the peeps and hoots that de Waal describes. In particular, this two syllable vocalization, with its distinct drop in pitch to the second syllable, contrasts sharply with the spectrogram of one of the de Waal food peeps, shown in Figure 6 (over).

Both vocalizations start at around 2000 Hz., but the de Waal food peep vocalization is a single syllable which is absolutely level in pitch.

The basic question remains, however: do Kanzi's vocalizations constitute English syllables? We address this question by proposing target phonemes. Operating by ear and by eye, we have started with the phoneme configuration of the English word 'onions' (predicted by the other strata) and lined these up with candidate segments of the waveform. Even where there appears to be no suitable candidate segment, we posit a target phoneme with the objective of finding out what is not sounded as well as what is sounded.

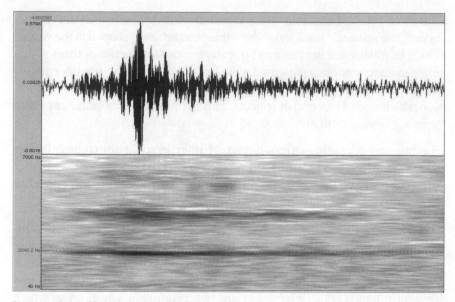

Figure 6: de Waal food peep spectrogram

The first thing we found was that there is no sound that is a possible candidate for /z/. This could be because Kanzi does not make a singular/plural distinction in lexicogrammar, or because he is unable to produce anything interpretable as /z/ because of his anatomy. These are questions to be explored as the methodology is applied to a large body of data. With that exception, we have been able to segment the vocalization so that there is one stretch to propose for each phoneme in *onion*.

The first syllable consists of two stretches, shown in Figure 7.

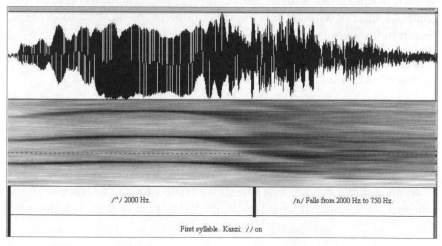

Figure 7: The first syllable of *onions*

The two proposed segments are clearly different. The first shows the sound energy distributed cleanly in three harmonics of 2000 Hz., 4000 Hz. and 6000 Hz. in the spectral range. The second segment is more complex, starting with the same three harmonics, but the harmonic energy is distributed much more broadly. The segment frequency drops sharply, stabilizing with harmonics of 750 Hz., 1500 Hz., and 2250 Hz. Throughout the entire second segment there is a broad distribution of energy across the frequency range of the spectrogram.

These differences between the segments are sufficient to force an interpreter to discriminate, to make a different choice for each within the English phoneme system. Given the probabilities established by the other strata, and the syllable structure of this stratum, it is not necessary to do more than re-cognize the first segment as the vowel /ʌ/ and the second as the nasal consonant /n/, a sequence which correlates with the choice of *onion* in the lexicogrammar, and with the syllable structure of Nucleus followed by Coda in the phonological stratum. There is no need for an interpreter to push the act of discrimination farther than acceptance of the target phonemes. It is not necessary to consider the actual sound more carefully to see if its elements correlate with different options in the phoneme system.

The second syllable consists of three stretches, shown in Figure 8.

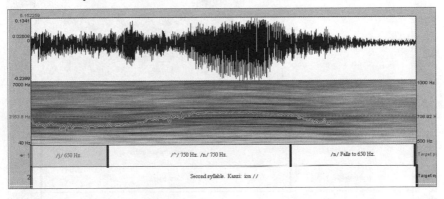

Figure 8: The second syllable of *onions*

The three proposed segments of the second syllable all contrast with those in the first syllable, having a fundamental frequency in the hundreds rather than in the thousands. But, more delicately, they differ from each other in terms of pitch movement. The first segment is level at 650 Hz.; the second rises to 750 Hz., where it levels off, and the third drops from there to 650 Hz. In terms of waveform pattern, these segments show the typical shape of a syllable: the first has relatively little amplitude, the second swells considerably, and the third drops back to the level of the first – and then tapers off to nothing. Given the probabilities established by the other strata, and the syllable structure of this stratum, it is again unnecessary to do more than re-cognize the first segment as the consonantal glide /j/, the second as the vowel /ʌ/, and the third as the nasal consonant /n/, a sequence which again correlates with the choice of *onion* in the lexicogrammar, and the with the syllable structure of Onset, Nucleus and Coda in the phonological stratum. There is again no need for an interpreter to push the act of discrimination farther than acceptance of the target phonemes.

3.3 The phonemes of *juice* in multistratal analysis

The bottom band of Figure 9 shows the context in an earlier phase of the interaction between Jared and Kanzi – again centered on food. The next band shows the interpersonal semantics: Jared makes an [offer], which Kanzi [accepts] through his vocalization of the word *juice* as well as through the physical act of drinking. The target syllable band shows //juice// as one syllable, for which we propose three target phonemes which are probable, given the choices at the other strata.

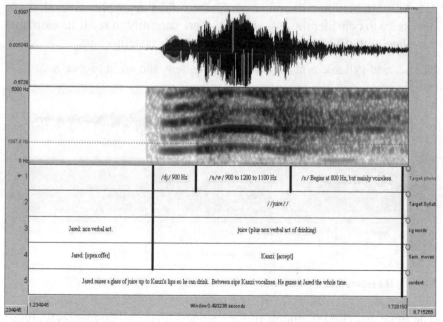

Figure 9: Kanzi accepts Jared's [offer] by drinking and vocalizing *juice*

The waveform shows a pattern consistent with a single syllable: the beginning has the low amplitude characteristic of a voiced Onset, and a frequency (900 Hz.) which is quite low for bonobo vocalization; the middle swells to the considerable amplitude characteristic of a Nucleus, with a rising-falling frequency (900 Hz. to 1200 Hz. to 1100 Hz.); and the end drops off in amplitude as would a Coda, starting with a level frequency of 800 Hz. and becoming voiceless. The choice of target phonemes correlates with choices made in the other strata, and their distribution here, /dj/ for the Onset, /u/ and /w/ for the Nucleus, and /s/ for the Coda, is in accord with English syllabification.

Given the probabilities established by choices in the other strata, and the syllable structure of this stratum, the discriminations made here are again minimal: the Onset is recognized as the consonant /dj/, the Nucleus as the vowel /u/ plus the vocalic glide /w/, and the Coda as the consonant /s/. There is no need to consider, for example, the extent of voicing in the Coda: a choice between /s/ and /z/ is relevant, and an 'easy' interpretation as /dj u w s/ has clearly been made by Jared, who says: '*yeah, we're having juice*', as shown in Figure 10 below.

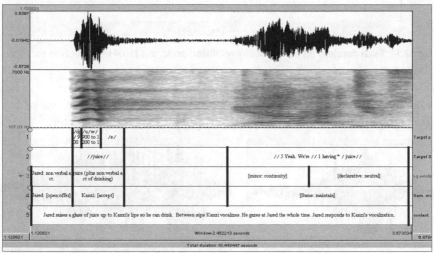

Figure 10: Jared interprets Kanzi's vocalization as *juice*

3.4 Finer phonetic discrimination forcing reinterpretation of the other strata

Figure 10 showed Jared interpreting Kanzi's first vocalization as *juice*, and Figure 11 shows Jared interpreting Kanzi's next vocalization in the same way, as *juice*.

On the basis of this interpretation of Kanzi's second vocalization, Jared's next move in discourse is a [maintain]. But, in fact, the two vocalizations are distinctly different. In the rapid flow of events, Jared had neither the time nor inclination to concentrate on Kanzi's sound and discriminate more delicately; he simply recognized the vocalization as consisting of the same phonemes he had recognized before – an interpretation which made perfectly good sense in the context.

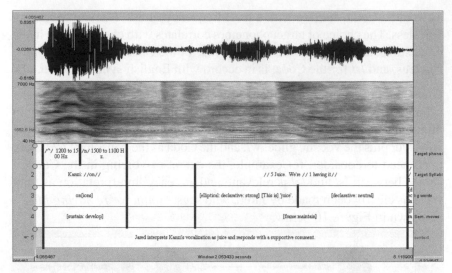

Figure 11: Kanzi develops his acceptance by vocalizing *onion*, but Jared interprets the vocalization as *juice*

Figure 12 shows differences between Kanzi's first two vocalizations which are sufficient to stimulate further discrimination by an interpreter.

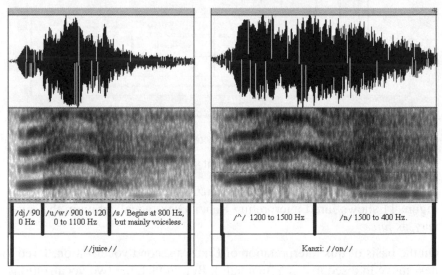

Figure 12: Kanzi's first two vocalizations compared

The waveform of the /s/ segment of the first image is characteristic of voiceless friction, while the waveform of the /n/ segment in the second image is like the /n/ segment in *onion* in Figure 7. It is particularly recognizable as /n/ because Kanzi has brought the frequency down to a range in which humans vocalize (750 Hz. in Figure 7 and 400 Hz. in Figure 12). Both the

/n/ segments have considerable amplitude, whereas the /s/ in *juice* does not. Difference between the two vocalizations is also evident when listening to them. The vowel qualities of the syllables are quite distinct. The first sounds more like a human /u/w/, and the second more like /^/. Also the sounds of the Coda phonemes are distinct, with the first sounding more voiceless, and the second distinctly nasal.

The conversation doesn't break down as a result of Jared's misinterpretation because his choices are not improbable, and, more importantly, because Kanzi is interested in pursuing his own agenda in a supportive way, not in confronting Jared (cf. Chapter 1). But, unlike Jared who must react within the tempo of the dialogue with Kanzi, we are able to listen to the sound as often as we wish, and our discriminations can take waveform and spectrogram information into account in our acts of iconic interpretation.

Having re-cognized different phonemes, we are forced to rethink the choices in the other strata in order to propose target phonemes which are in harmony with the newly discriminated sounds. In the context Kanzi's fondness for onions is relevant. One of the interpersonal MOVE choices available to him, [sustain: develop], allows Kanzi to stay within the general acceptance of Jared's provision of food, while extending it to include onions as well as juice. *Onion*, therefore, is a probable word, and the target phonemes of its first syllable, /^/ and /n/, are distributed as they were in the first syllable of onion in Figure 7. There are many occasions in discourse when such careful discrimination of the sound affects choices in the other strata, as when, for example, the speaker has a speech impediment, or talks with a strong accent, or a hearer is hearing impaired. In such situations, even considerable concentration on the sound will not guarantee correct interpretation.

Jared, in fact, continues with discourse based on his *juice* interpretation, in the face of which Kanzi produces another vocalization which is more easily recognized as *on(ion)*, using the [append] choice in the MOVE system to jump over Jared's communication and repeat his earlier development of the food options to include onions as well as juice.

Figure 13 (over) shows that this third vocalization is similar to the immediately previous one and to the first syllable of *onion* in Figure 7.

Figure 13 shows two clearly distinct segments. We have aligned them with the target phonemes /^/ and /n/ for the same reasons that we did so in the discussion of Figure 12. Kanzi's third /n/ vocalization resembles both his second /n/ vocalization and the /n/ in the first syllable of *onion* in Figure 7. Kanzi's third /n/ vocalization, at 300 Hz., is even farther down in the human range. None of these three vocalizations resembles the one-syllable *juice* discussed in Figure 9. In all of the /^/n/ syllables, there are clear harmonics in the first phoneme, with much more widely distributed energy in the second. But all three /n/ segments show a flat, strong first harmonic which is very low for a bonobo, as Kanzi brings his pitch down in producing a sound more readily interpretable by humans. When heard as isolated segments, all three have a distinctly nasal quality absent from /s/ of *juice*.

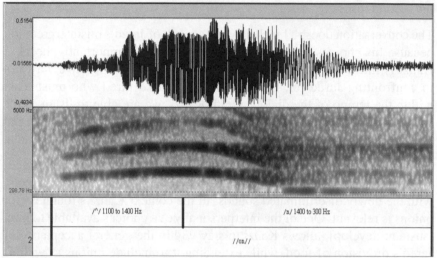

Figure 13: Kanzi's third vocalization

In this part of the interaction with Jared, Kanzi vocalized three times.

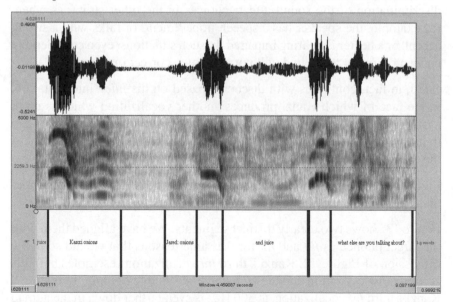

Figure 14: Jared's interpretation of Kanzi's *onion* vocalization as *onion*

A few minutes later, Kanzi repeats his vocalization of *onion*, but this time with the second syllable (Figure 5 above). On this occasion, Jared clearly interprets Kanzi's vocalization as *onion*, as shown in Figure 14.

4   Conclusion

In this chapter, we have examined a few seconds of sound, and have proposed a small number of target phonemes for a few bonobo vocalizations. The methodology, however, is designed for application to a large body of data – extensive records of discourse interactions between apes and humans, with the goals of:

(1) specifying in much greater detail the phonetic parameters which correlate with particular phonemes;

(2) proposing a table of the English phonemes which Kanzi is able to produce;

(3) examining formant patterns in those parts of Kanzi's vocalizations which are low enough in frequency;

(4) demonstrating that language competent bonobos vocalize in a dialect of English.

References

Deacon, T. W. (1997) *The Symbolic Species: the co-evolution of language and the brain*. London and New York: W. W. Norton.

de Waal, F. (1988) The communicative repertoire of captive bonobos (*Pan paniscus*) compared to that of chimpanzees. *Behaviour* 106: 183–251.

Edelman, G. and Tononi, G. (2000) *A Universe of Consciousness: how matter becomes imagination*. New York: Basic Books.

Hopkins, W. D. and Savage-Rumbaugh, S. (1991) Vocal communication as a function of differential rearing experiences in *Pan paniscus*: a preliminary report. *International Journal of Primatology* 12(6): 559–83.

Savage-Rumbaugh, S., Fields, W. and Spircu, T. (2004) The emergence of knapping and vocal expression embedded in a pan/homo culture. *Biology and Philosophy* 19(4): 541–75.

Taglialatela, J., Savage-Rumbaugh, S. and Baker, L. (2003) Vocal production by a language-competent *Pan paniscus International Journal of Primatology* 24(1): 1–17.

Taglialatela, J., Savage-Rumbaugh, S., Rumbaugh, D. M., Benson, J. and Greaves, W. (2004) Language, apes, and meaning-making. In G. Williams and A. Lukin (eds) *The Development of Language: functional perspectives on species and individuals*. London: Continuum.

Index